U0265336

当代建筑师系列

胡越

HU YUE

胡越工作室　编著

中国建筑工业出版社

图书在版编目(CIP)数据

胡越/胡越工作室编著.—北京：中国建筑工业出版社，2012.6
(当代建筑师系列)
ISBN 978-7-112-14276-7

Ⅰ.①胡… Ⅱ.①胡… Ⅲ.①建筑设计－作品集－中国－现代②建筑艺术－作品－评论－中国－现代 Ⅳ.①TU206 ②TU-862

中国版本图书馆CIP数据核字（2012）第085862号

整体策划：陆新之
责任编辑：徐 冉 刘 丹
责任设计：董建平
责任校对：刘梦然 陈晶晶
版面设计：卢 超

感谢山东金晶科技股份有限公司大力支持

当代建筑师系列
胡越
胡越工作室 编著
*
中国建筑工业出版社出版、发行（北京西郊百万庄）
各地新华书店、建筑书店经销
北京嘉泰利德公司制版
北京顺诚彩色印刷有限公司印刷
*
开本：965×1270毫米 1/16 印张：11 字数：308千字
2012年8月第一版 2012年8月第一次印刷
定价：98.00元
ISBN 978-7-112-14276-7
(22345)

目 录　Contents

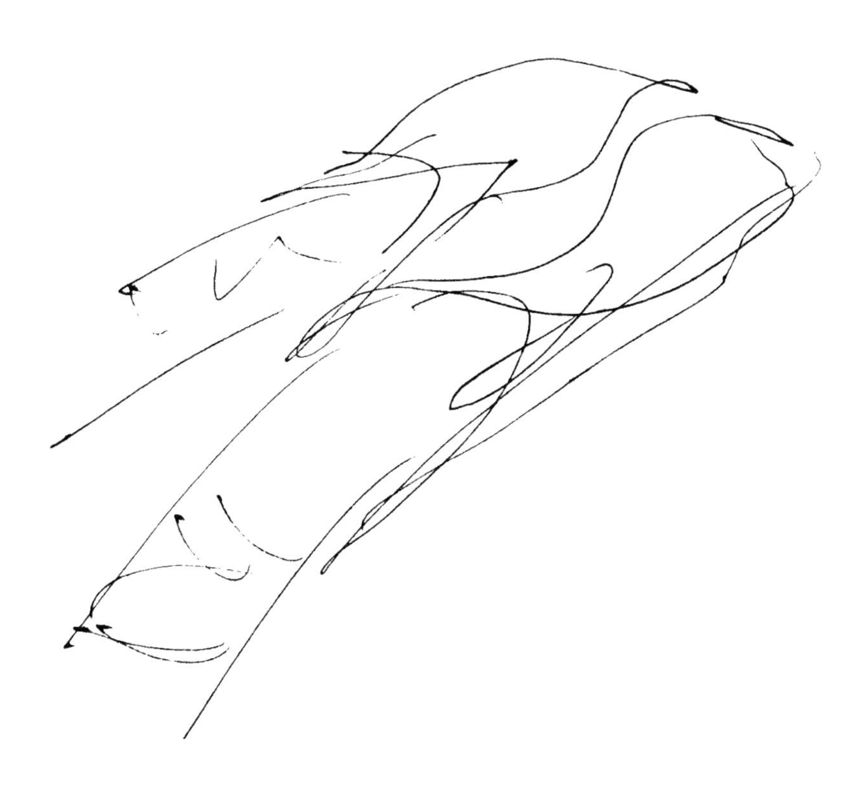

胡越印象

文／黄元炤

胡越，1964 年出生，1986 年于北京建筑工程学院毕业后，直接进入到北京市建筑设计研究院工作至今，之后成为院总建筑师和全国建筑设计大师。2003 年，胡越工作室成立，是北京院第一批并在 2005 年的改革中成为院属独立工作室。在成立工作室之余，胡越的创作步伐不曾减缓，并持续性地爆发，从对材料的研究到关注城市空间，再到关注方法与编写任务书，他努力发现独特的观点，通过观点来创作出一系列独特的建筑。

在北京院工作期间，胡越参与了不少的项目创作。北京国际金融大厦，是他于 20 世纪 90 年代的代表性作品，也使他获得了许多的奖项与业界的名声。这个项目是个尊重当地城市既有街廓与肌理的设计，建筑恰如其分地以规矩方整的体块出现，而体块之间也形成虚实扣合的形体，形体的巨大化，给人一种霸气之感，这就有点当时 KPF 建筑师事务所的风格。由于金融大厦是个金融机构，胡越特别注重办公与商业功能的考虑。除此之外，在立面上，他用现代的建筑材料——铝合金窗式玻璃幕墙，去表现出传统民族图案的组合，他企图在允许局部装饰的背后，表述出传统在现代基础上寄居的状态。这样的语言与手法异于他早期设计英东游泳馆双层屋面时，所表现的既具象又抽象的符号象征。所以，在北京国际金融大厦项目中，胡越在功能的基础上，似乎从关注大的形式与符号转为关注小的局部与细节，而细节上民族图案幕墙的设计，一方面是为了符合当时北京市高层的政策，另一方面也是他受到努维尔设计的巴黎阿拉伯中心的影响。但在双重影响下，胡越其实是不太赞成传统的符号、隐喻、暗示的表述与地域性的设计倾向的，有时是迫于现实的无奈而必须表现出这样的设计。

对材料与构造的关注与研究，是胡越早期设计时所取重的一点。在北京国际金融大厦中，他采用窝式幕墙系统，分解成 4 米 ×3.6 米单元，用传统方法将其固定在钢框架中，这是他对材料的初步尝试。之后 20 世纪 90 年代末，他出国探访与游历，看了许多建筑，并作了一些思考，同时翻阅不少建筑杂志，这段期间成为他在思想上的重大转折点。当时胡越经思考后，觉得普遍的中国建筑师对于材料与构造是不了解的，且处于一种被动的状态，没有把材料与构造当作是创作的动力。由于这样的激荡与反思，他的视点开始经历从一个巨观到微观，与从一个整体到细部的转变，更多地关注材料问题。当时对玻璃感兴趣，有意识地去做研究与探索，之后验证在设计创作当中。望京科技园二期，就是胡越对于玻璃幕墙研究后的成果展现，他根据体形的变化与功能的要求设计 4 种玻璃幕墙，有隐框单元式玻璃幕墙、密肋式玻璃幕墙、显框分格渐变式玻璃幕墙及双层通道式玻璃幕墙，且分别选用透明玻璃和印刷玻璃，建筑因玻璃材料的使用而使几何体量之间的关系与对位更加清晰。

望京科技园二期，除了玻璃材料的使用外，在建筑语言上，胡越倾向于一种极少性的设计表述。他将建筑还原到最原初的状态，只表现简单的几何体量的构成。建筑顶端，胡越设计了大悬挑与出挑的体量，在当时中国建筑界还没有类似的设计出现，这是受到了当时先进的建筑思潮的影响，可见胡越是个追随潮流，且跟风比较紧的建筑师。他企图在体量的厚重当中，去追求单一几何构成的形态展现，加上玻璃的纯粹运用，让建筑展现现代时尚之感。胡越似乎想从设计中，追寻一种建筑在现实生活与环境中，升华到某种单纯力量的表述，体现的是一个简单，且具有时尚品格与质量的建筑。

胡越，是以单纯研究、试验与开发材料的视点切入设计，并逐渐带出建筑的走向，材料的选择是影响他的作品最终的表现形式。胡越之于玻璃、金属网、聚碳酸酯板材、铝合金方管、钢筋混凝土，就如同安藤忠雄之于清水

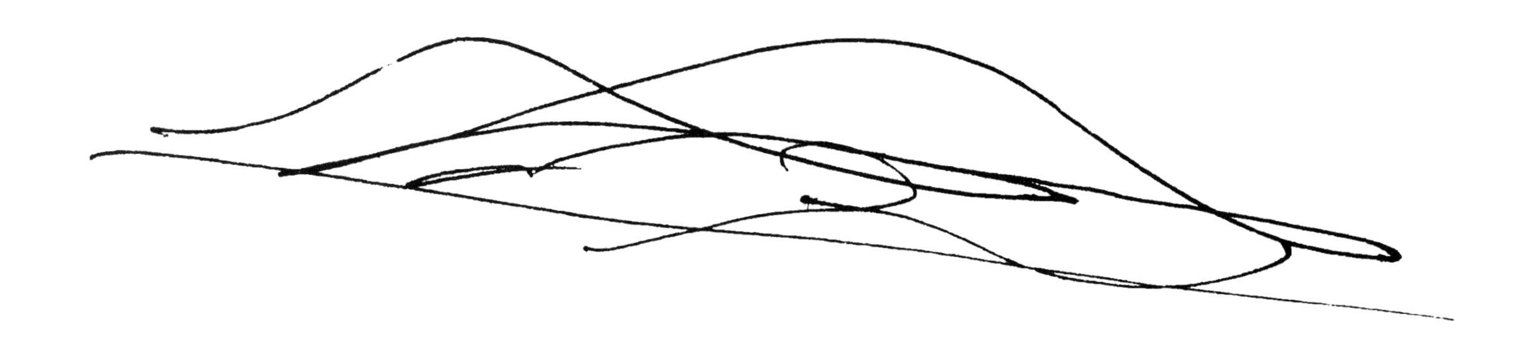

混凝土，坂茂之于纸，远藤秀平之于波形钢板，所以，胡越是个以材料取重的建筑师。也因为胡越关注材料与构造的倾向，他开始尝试与创造出了一条偏向于表象的设计路线，即是在既有结构体外新增一层皮，但不单单只是皮层的体现，有时皮层也会带出与内部空间相互的联系与关系，或者形成一个皮层的造型体。在五棵松棒球场中，在深灰色建筑外墙外，局部罩一层疏密相间的金属网外皮。在上海青浦体育馆与训练馆改造中，同时用三种材料相互展现，中间以聚碳酸酯板材单层的方式，上层是纵与横向的编织手法，围绕在建筑物的外层，下层是铝合金的穿孔钢板与铝合金方管的组配。在上海世博会 UBPA 办公楼中，在各个立面采用均质开窗的形式，通过外墙涂料颜色的变化、拼接以及由于功能需求形成了不同材质（织物膜、铝合金）的体量穿插。

北京建工学院新校区六号楼，是胡越近期的新作品，他似乎想从这个项目中跳开他之前尝试皮层设计的外在印象，转到关注功能与结构。为了使建筑具有更大的使用性与灵活性，他将建筑设计成一层，这是功能性考虑，也因为是一层，他把建筑做成无柱，且不用钢结构，用钢筋混凝土建成。这个项目体现出功能与结构之间的巧妙结合，为了创造出大的空间，而去思考结构的应用与适宜性。

虽说胡越的作品都带有点表皮的设计倾向，但他自己并不会刻意去突显这一点。他的作品与作品之间的差异都很大，从早期的北京国际金融大厦，到望京科技园二期，到上海青浦体育馆与训练馆改造，再到北京建工学院新校区六号楼，在场地条件、材料的使用、功能的考虑、策略的拟订等方面，他均会根据每一个项目采取不同的措施，根据具体的状态去想设计的过程与结果。另外，也可从作品的差异当中，看到胡越似乎不想把自己定性。当每一个项目建成后，他都想要赶紧跳脱开来，因为他觉得设计应该与时俱进，所以，不应该墨守成规，只选择做一样事。他更觉得设计应该要跟上潮流，要缩短与最先进文化之间的差距，应该关注时尚与未来，他认为建筑都是在设计未来。所以，他总是在当下的时空中，去想到设计的未来。胡越的建筑态度始终处于一个进行时与未来时，而不是过去时。

方法论，是胡越的博士论文主攻的研究领域，更是他目前最关注的一点。他企图从对方法的钻研与琢磨中，梳理出自己对设计的一个切入点，而原来他所关注的材料，已只是他设计中的一部分。胡越想让他的方法论朝向探索未来的发展趋势与社会的需求方向，他也认为方法是建立在最开始的需求上，是为了解决问题，才有了方法。在他所关心的项目进行前，他会编写任务书，而在编写中，他想去发现建筑中独特的问题，并以此作为创作的重点，从中发现独特的观点，通过这些观点表现在建筑上，让他的建筑具有一种独特性，有他自己的价值判断与哲学视角。

综观胡越的作品都有着强烈的视觉性，除了材料的表述外，也有鲜明的色彩展现，如望京科技园二期的幕墙上的蓝、银相间，上海青浦体育馆与训练馆的白、灰、黑、黄的各自存在，上海世博会 UBPA 办公楼外墙上的深灰、淡灰、红的活泼混搭等，在色彩上，给人一种很强烈的个性化与风格感。但是最终，风格并不是胡越所要追求的，他觉得建筑，不管是室内或室外，必须给人感觉舒服。他更追求室外给人舒服感的营造，因人都在建筑和别的建筑所构成的城市空间中穿行着，而建筑师就是城市空间的设计者，有着一份责任。所以，他更愿意去创造室外空间的舒服与轻松的状态，这是他的一个终极目标，也是他的设计追求。

Portrait

By Huang Yuanzhao

Hu Yue, born in 1964, graduated from Beijing University of Civil Engineering and Architecture in 1986, has been working in Beijing Institute of Architectural Design since then. Now he is the chief architect of the Institute and the National Architectural Design Master. In 2003, Huyue Studio was founded, which was the first group of its kind in China, and the Studio became independent to the Institute with the reform policy in 2005. After his studio was established, Hu Yue has not slowed down his pace of architecture creation, and he has been making successive achievements in his field. From materials research to city space concern, and then to focusing on architecture methods and Design Brief, he tries to develop his unique point of view, and through such view to create a series of unique buildings.

In Beijing Institute of Architectural Design, Hu Yue has participated in many projects creation. Beijing International Finance Building was his representative works in 1990s, as a result he won various awards and reputation in this field. This project is designed to respect both local city's existing contour and texture, the building is constructed in the form of proper square blocks, and virtual and real buckled body are formed between blocks, and the building's huge body impresses people with a domineering sense, so this is some kind of KPF architecture firm. Since the Finance Building serves as a financial institution, Hu Yue takes its office and business functions into special consideration. In addition, he applies modern building materials in facade and aluminum-alloy-window-type glass curtain wall is used to show the design portfolio of traditional ethnic patterns. He attempts to express the tradition's lodging on a modern basis with partial decoration, such architecture language and method are quite different from his earlier design style of the double roof of Ying Tung Natatorium which demonstrates concrete and abstract symbols. So, in Beijing International Finance Building project, Hu Yue seems to transfer his focus from large forms and symbols to small local parts and details on the functional basis. His curtain wall design of ethnic patterns is to be in accordance with Beijing authority's policy at that time on one hand, and on the other hand he has been influenced by Paris Arabian Center design by Nouvel. But under such dual influence, Hu Yue is actually not much in favor of expressions of traditional symbol, metaphor, hint and regional design tendency, sometimes under pressure of helpless reality he has to make this type of design.

Interest in and study on materials and structure was Hu Yue's early design focus. In Beijing International Finance Building project, he adopted a nest type curtain wall system with 4mx3.6m decomposition units, and applied traditional methods to fix units on steel frame, which was his initial attempt to use such material. At the late 1990s, he went abroad for visit, observed many buildings and had some thinking. Furthermore he read various architecture magazines, so this was a great turning point in his architecture thinking. At that time Hu Yue believed that Chinese architects knew little about materials and structures.They were passive, and did not take material and structure as the motive power for creation. So after his thinking and reflection his architecture views changed from macroscopic to microscopic, and from the whole to details. He began to attach his attention to material and felt interested in glass, and conducted some research and exploration of glass, and then used and tested such material in his design creation. Wangjing Science and Technology Park (Phase II) is Hu Yue's achievement displaying of his study on glass curtain wall. According to changes in size and function demand he created 4 kinds of glass curtain wall: concealed frame unit glass curtail wall, compact rib glass curtain wall, displayed frame and gradual-changed glass curtail wall, and double deck channel glass curtain wall with transparent glass and printed glass. Glass makes the building's relationship among geometric volumes and their counterpoints more clear.

In addition to use of glass material in Wangjing Science and Technology Park (Phase II), Hu Yue tended to adopt a minor design expression in architectural language. He reverted the building to its original state, only to express formation of simple geometric volumes. At the building top, Hu Yue designed a large jettied and outrigger volumes. At that time there was no similar design in China's construction industry, and Hu Yue was influenced by the advanced architecture ideological trend. Hu Yue is such an architect who greets trend and follows it closely. He attempts to pursue a single geometric shape displaying in volume, thickness and weight. With pure use of glass, the building tends to be modern and fashionable. In architecture design, Hu Yue seems to pursue an expression that architecture distillates itself to be a pure power in real life and environment so that it is able to embody it a simple building but with a fashion taste and quality.

Hu Yue makes design from a perspective of pure research, testing and developing materials, and gradually he leads the trend of construction, and material selection influences his works ultimately. Glass, wire netting, polycarbonate sheet, aluminum alloy square tube, reinforced concrete to Hu Yue is like the plain concrete to Tadao Ando, paper to

Shigeru Ban, and corrugated metal panel to Shuhei Endo. Therefore, Hu Yue is an architect who attaches special attention to material. Since Hu Yue focuses his attention on the surface of materials and structural, he attempts to develop an appearance-bias design line, namely on the external surface of existing structure a new layer is added, but it is not only a surface embodiment.Sometimes the surface also demonstrates a mutual connection and relationship with internal space, or forms a surface-building body. In Wukesong Baseball Center, on the dark grey exterior wall, well-dense metal sheath is surfaced. In the renovation of Shanghai Qingpu Stadium and Training Hall, three kinds of materials are presented in a single layer of polycarbonate sheet with an upper layer made with longitudinal and transverse weaving techniques over the external surface of the buildings and a lower layer made with aluminum alloy perforated plates and aluminum alloy square tubes. In Shanghai UBPA Office Building for the World Expo, homogeneous fenestration is used in each facade, different materials (fabric film and aluminum alloy) volumes are interspersed with exterior paint color changes, splicing and due to functional requirement.

Building No. 6 of New Campus of Beijing University of Civil Engineering and Architecture is Hu Yue's recent new work. He seems to want to jump out of surface-designed impression in this building that he tried before, so he turned to focus on function and structure. In order to invest the building with more use and flexibility, he designed the building with one floor, and such design was out of functional consideration. Since this was one floor building, it has no columns without steel structure, and it was made of reinforced concrete. This project reflects ingenious combination of function and structure. In order to create a large space, structure application and suitability are taken into deep consideration.

Although Hu Yue's works have shown some surface design tendency, he does not deliberately highlight this characteristic. There is big difference among his works. From the early Beijing International Finance Building and Wangjing Science and Technology Park (Phase II) to Shanghai Qingpu Stadium and Training Hall and then to Building No. 6 of New Campus of Beijing University of Civil Engineering and Architecture, in aspects of site condition and coping, use of materials, functional considerations and strategy making, he takes different measures according to various projects. According to specific condition he figures out design process and results. In addition, from the works' differences, it can be discovered that Hu Yue seems not to position himself to certain design style. Once a project is completed, he wants to break with its design influence, because he thinks that the design shall develop with times. Therefore, as an architect he should not confine himself to one style. When he just chooses to make a design, he should follow design tendency to shorten his gap with the most advanced culture, and pay more attention to fashion and future.From his point of view, constructing building is designing future. Therefore, in the present time, he always thinks of the design future, his architectural attitude is the progressive and future tense, not the past tense.

Methodology is Hu Yue's doctoral thesis main research field and the most important point that he concerns with. He attempts to figure out a penetration point of designing from his method research and analysis. Material that he showed his concern with before is only a part of his design. Hu Yue wants to develop his methodology towards exploring the future development trend and the social demand. Method is based on initial demand, and solution is for solving problem. Therefore, he is trying to detect problems. Before a project begins, he writes a task statement himself, and in preparation of the task statement, he wants to find unique problems about architecture.He takes those problems as the key points for his creation to get unique perspective, and makes these views expressed in architecture, so he is able to make his architecture as unique and he can make his own value judgment and philosophical perspective.

All Hu Yue's works produce strong visual effects.In addition to material expression, there is bright color expression.For example, in curtain wall of Wangjing Science and Technology Park (Phase II), same blue and silver colors are well-dense; in Shanghai Qingpu Stadium and Training Hall, white, grey, black and yellow are showing respectively; in the exterior wall of Shanghai UBPA Office Building for the World Expo, dark,gray, grayish and red colors are lively mixed to present people with a very strong sense of individualization and style. But style is not what Hu Yue is pursuing. He thinks that building, either interior or exterior, must make people feel comfortable, and he tries to create a sense of comfort out of exterior buildings, because people are moving in the city that are made of buildings, and the architect is the designer behind the city space and has a sense of responsibility, so Hu Yue prefers to create comfortable and relaxed outdoor space,which is one of his ultimate goals and his design pursuit.

北京国际金融大厦　　北京

Beijing International Finance Building, Beijing

1998

北京国际金融大厦是由招商局全资公司北京金龙兴业房地产有限公司开发的大型房地产项目。大厦主要以金融机构办公、营业为主。

北京国际金融大厦位于北京西长安街南侧，距天安门广场约 3 公里。它西临远洋大厦、电教中心，北面与长话大楼、中国工商银行总行相对，处在北京金融街开发区的最南端。

大厦由四个相对独立的办公楼和两个弧形连接体组成。北侧为 11 层，南侧 14 层，四幢办公楼首层为银行营业厅，北侧二至十一层，南侧二至十三层为办公楼，南侧十四层为餐厅和设备机房。大厦地下两层，一层为金库、账库、保管库、车库、自行车库及快餐厅；二层为机房和车库。

北京国际金融大厦占地约 1.74 万平方米，建筑面积约 10 万平方米。1996 年 6 月完成初步设计，同年 9 月完成施工图。工程于 1998 年 6 月基本完工交付使用。

在建筑设计中我们主要遵循三个原则：1. 与原有城市结构相协调；2. 满足业主在地产开发和管理上的要求；3. 创造与整体环境协调又富于个性的建筑。为了与周围新老建筑在体型上取得平衡，大厦采用了长 134.9 米、宽度 69.7 米、高 45 米构图完整的大体型。同时为了减轻道路南侧大体型建筑对长安街的压迫感，并考虑到面积和容积率的限制，采用了化整为零的手法，将巨大的体型分解成三大部分，即中央大厅及尖顶、四个办公楼、两个巨大的弧形连接体。这样就解决了城市环境和建筑规模之间的矛盾。

内景 / Interior view

由于长安街一线规划对建筑高度有严格的限制，因此在整齐划一的体型下，运用前后两排建筑之间的高度变化，丰富了建筑造型，增加了层次，同时利用透视的原理，减轻了建筑高度对街道的影响。

为突出建筑的个性，同时结合建筑功能上的要求，利用屋顶机房、水箱间等设施，在长方形的体形中央部分设计了四个标志塔。这四个标志塔将中央大厅与四个办公楼紧密地联系在一起，在横向的立面构图中加入了竖向构图元素并强调了建筑中轴线。这样使大厦在其所在区域具有很强的可识别性。

在建筑顶层用两个巨大的弧形连接体，将四个办公大楼两两相连，流畅的曲线打破了长方形构图的呆板，使整个建筑活泼而富有生气。

大厦是由四个长方形办公楼组成的。办公楼的布置适应了业主在商业上的需求，同时也有利于物业管理。在四个长方形办公楼的底层是四个银行营业厅，它们被位于大厦中心的一个圆形大厅组织在一起，大厅内部高 10 米，中央有一个钻石形锥顶。透过玻璃锥顶向上是由四个办公楼、四个标志塔、两个巨大的门洞和弧形连接体组成的，在蓝天映衬下的丰富壮观的建筑空间。这样就在中央大厅内有限的、较经济的空间中获得了巨大的、无限的空间感受，满足了银行大厅对建筑空间的要求，同时在造型上和空间上获得非常富有个性的特殊效果。

大厦外装修采用了铝板、玻璃幕墙和石材。在立面构图中占主要地位的玻璃幕墙采用了现代建筑材料与传统图案相结合的手法，试图创作一个既体现时代精神，又具有民族特色的新建筑。结合国情和考虑了造价等因素，我们设计了独特的窗式幕墙系统。该系统将幕墙分解成 4 米 x3.6 米的单元，采用传统方法将其固定在钢框架中，既方便施工又节省资金。在中央大厅采用了点式连接玻璃幕墙，获得了良好的效果。

总平面图 / Master plan
内景 / Interior view

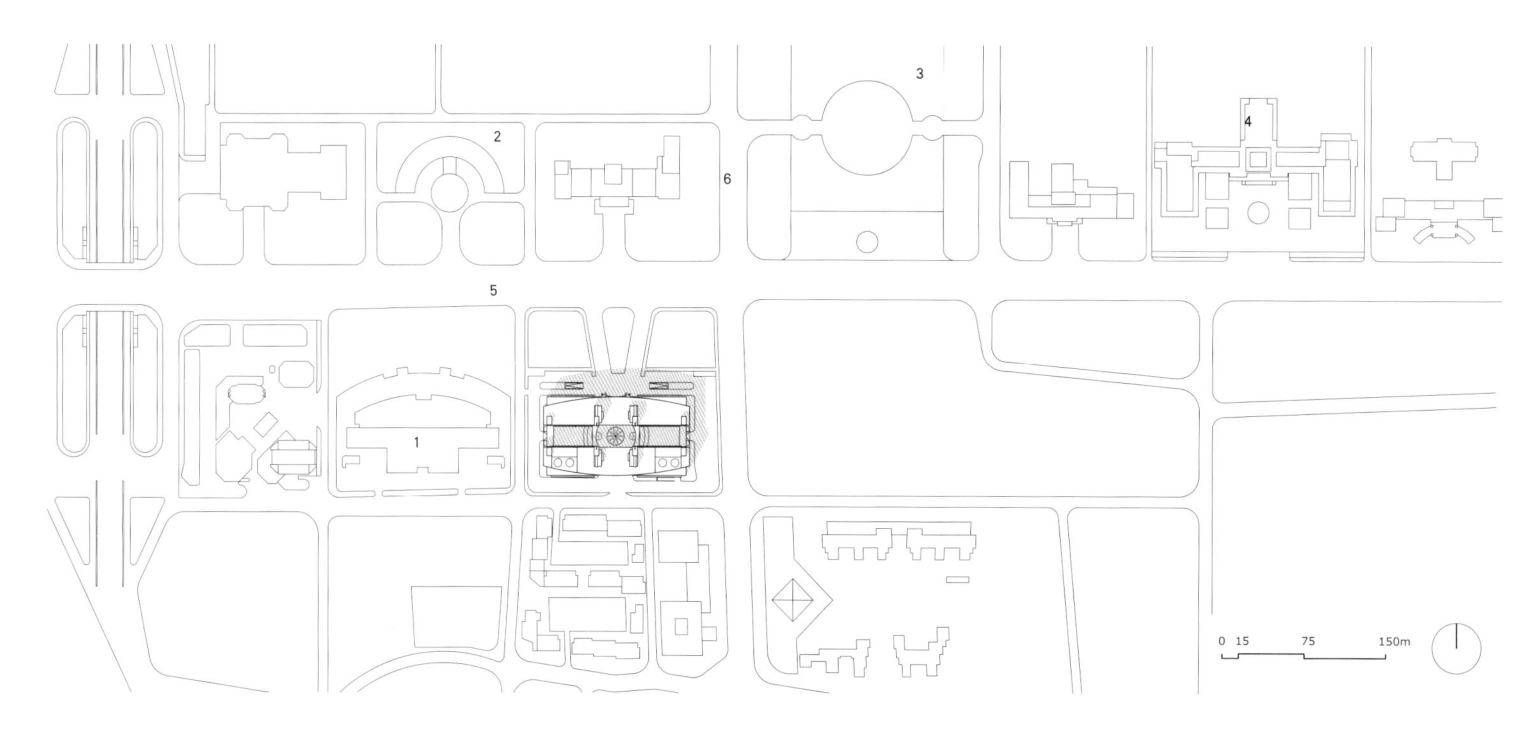

1 北京远洋大厦 / Beijing Cosco Building
2 中国人民银行 / The People's Bank of China
3 中国工商银行 / Industrial and Commercial Bank of China
4 民族宫 / The Cultural Palace of Nationalities
5 复兴门内大街 / Fuxingmennei Street
6 闹市口中街 / Naoshikou Middle Street

Beijing International Finance Building is a large-scale real estate project developed by Beijing Dragon Atates Properties Co., Ltd, a subsidiary wholly-owned by China Merchants Group. The mansion is mainly for offices and business operation of financial institutions.

Beijing International Finance Mansion is located on the south side of West Changan Street of Beijing and about 3 km to the Tiananmen Square. It is west to Cosco Plaza and E-education Center, opposite to Beijing Long Distance Call Building and ICBC HQ in the north and at the south tip of Beijing Financial Street Development.

The mansion consists of four relatively separate office buildings and two arc connection structures, arranged with 11 floors on the north side and 14 floors on the south side. The four office buildings have the first floor arranged for bank lobby, the 2nd - 11th floors on the north and the 2nd -13th floors on the south for offices and the 14th floor on the south for restaurant and equipment room. The mansion has two basements (B1 and B2), of which B1 is meant for treasury, account storage, vault, garage, bike parking area and fast food restaurant and B2 for machinery room and garage.

外景 / Exterior view

首层平面 / The 1st floor plan
标准层平面图 / Plan for standard floor
外景 / Exterior view

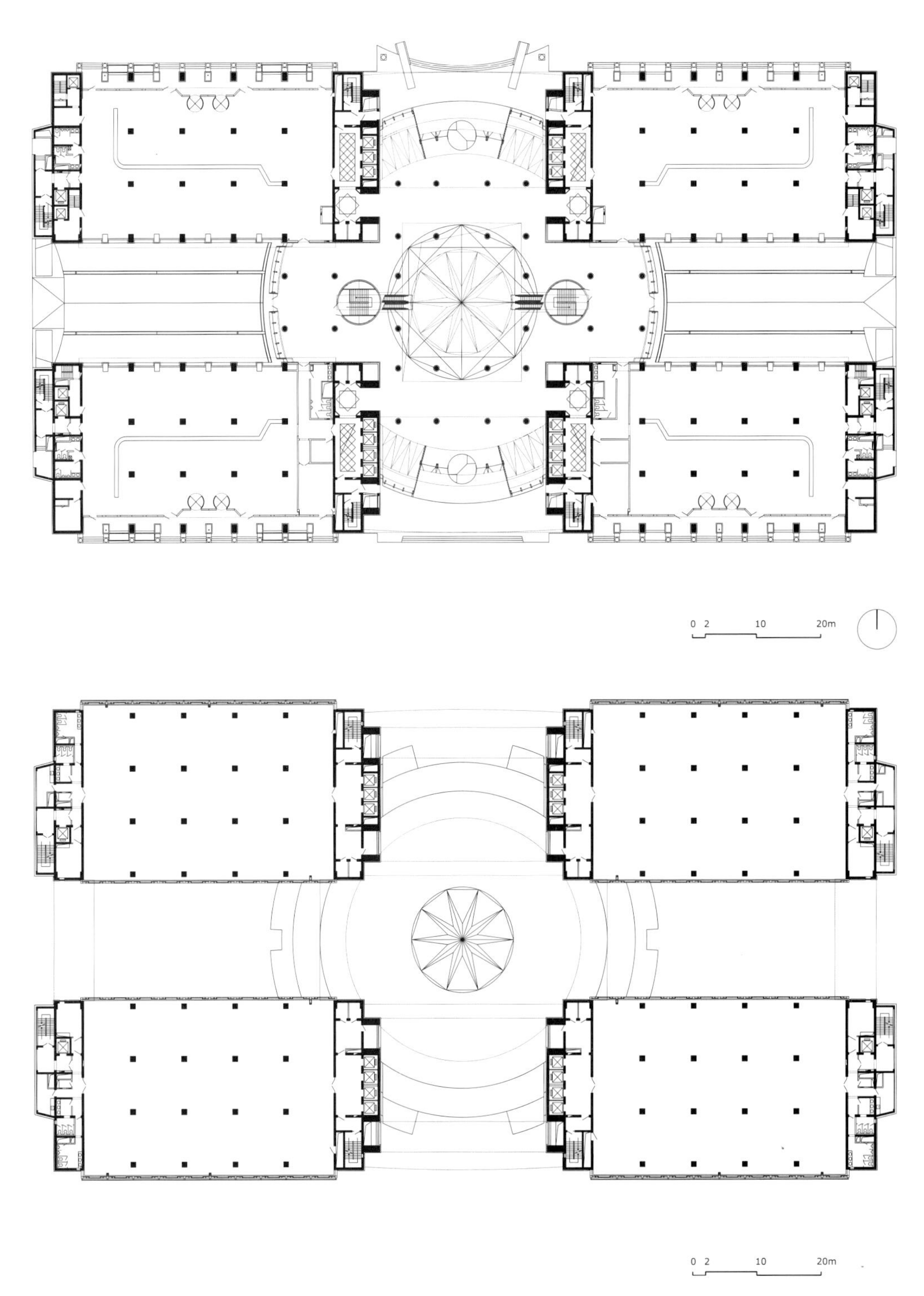

Beijing International Finance Mansion has a land area of 17,400m^2 and a floorage of about 100,000m^2. We completed the primary design in June ,1996 and construction drawings in September of the same year. The project was substantially completed for delivery in June 1998. During the architectural design, we followed three principles: 1) coordinating with the existing urban structure; 2) meeting the requirements of the Employer for real estate development and management; 3) creating a building harmonized with the overall environment and full of personality.

To balance with the surrounding new and old buildings by size, the large size of 134.9m long, 69.7m wide and 45m high with an integral pattern was used for the mansion. Additionally, to reduce the pressure of large-size building on the south side of the road on Changan Street and with consideration to the limitation of area and capacity, the method of breaking up the whole into parts was used to decompose the huge size into three parts: central lobby and lancet, four office buildings, and two huge arc connection structures. In this way, the conflict was resolved between the urban environment and the building scale.

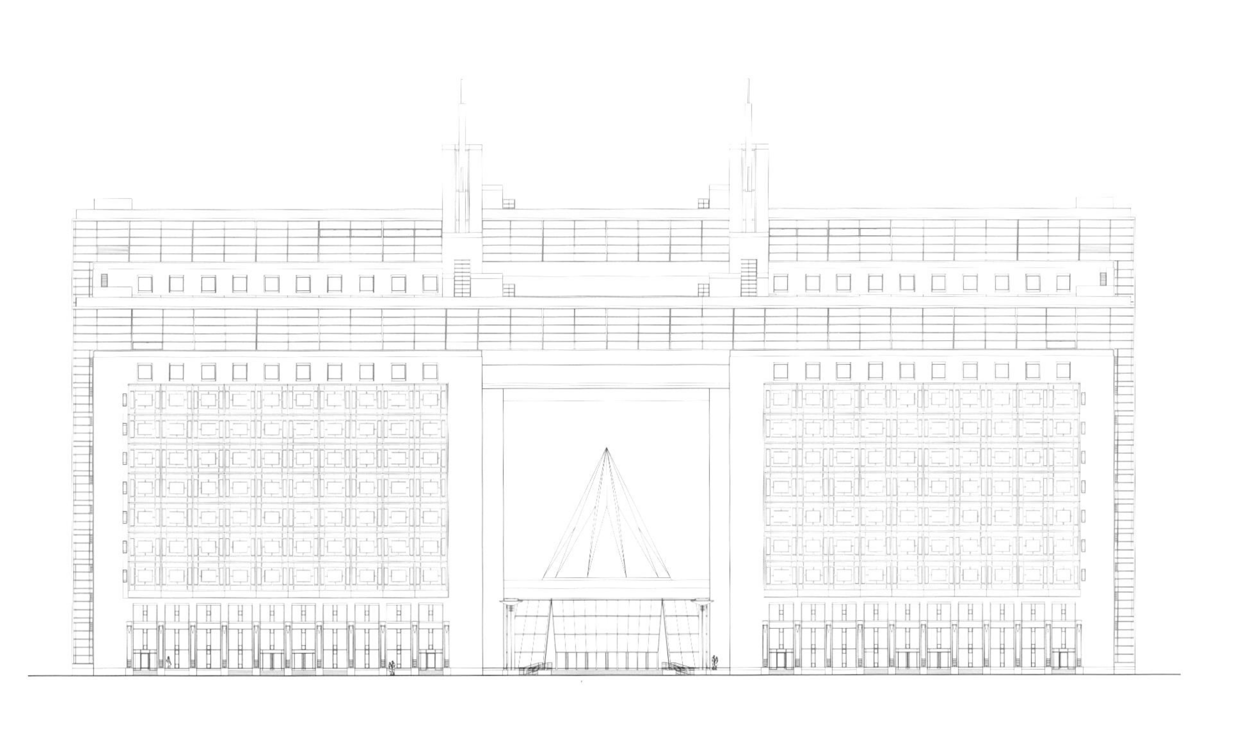

外景 / Exterior view
立面 / Elevation

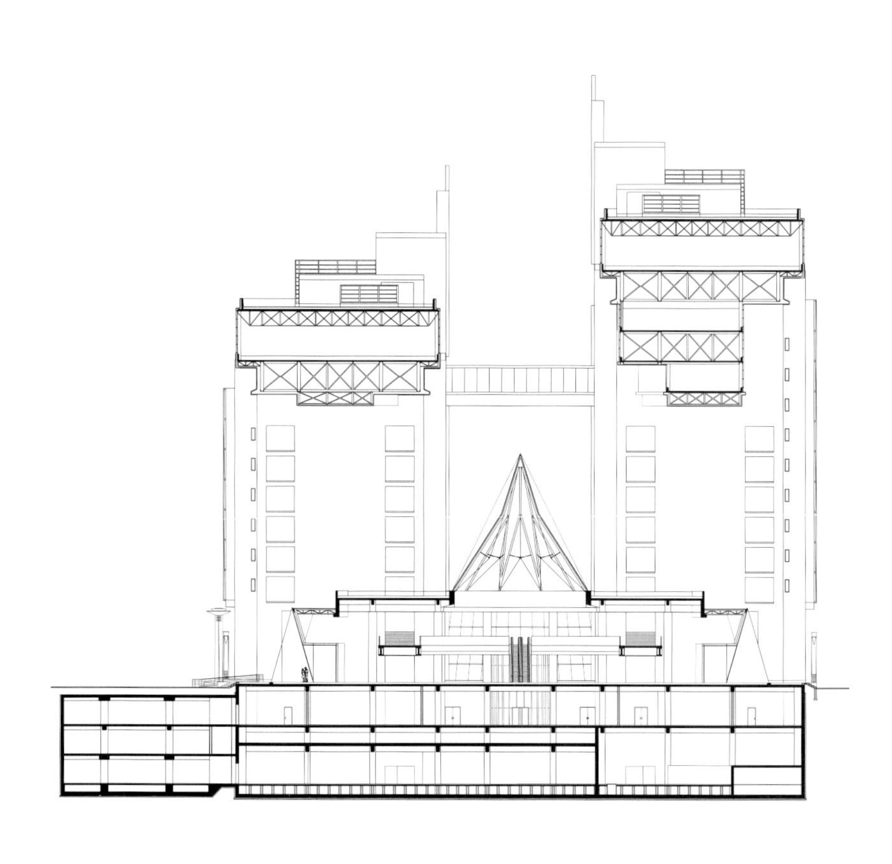

细部 / Details
外景 / Exterior view
剖面 / Section

Since the planning along Changan Street structures provides strict restriction for height, the application of height variation between front and rear rows of buildings with a neat and integrated size has enriched the architectural modeling and increased the hierarchy, and the perspective principle was adopted to reduce the impact of the building height to the street.

To highlight the personality of the buildings as well as to comply with their functional requirements, by making use of such facilities as roof machinery room and water tank room, four symbolic towers were designed in the central part of the rectangle size for connecting the central lobby closely with the four office buildings. By adding the vertical pattern elements in the transverse elevation pattern and highlighting the central axis of the building, the mansion became identifiable in the local region.

On the top of the building, two huge arc connection structures were used to link the four office buildings in pairs, while the smooth curve broke the rigidity of the rectangle pattern to make the whole architecture more vivid and vigorous. The mansion consists of four rectangle office buildings which have been laid out to meet the commercial demands of the Employer as well as to benefit the business management. At the basement of the four rectangle-type office buildings

are four bank lobbies that are organized by the round lobby located in the center of the mansion. The lobby is 10m high internally, with one diamond shaped cone in the center. Upward through the glass cone is the abundant and grand architectural space under the blue sky, which was formed by the four office buildings, four symbolic towers, two huge gateways and arc connection structures. Thus, a huge and infinite space experience was created in the relatively limited and moderate central lobby, meeting the architecture requirement of the bank lobbies as well as gaining a special effect with rich personalities in forms and space.

The mansion is externally finished with aluminum boards, glass curtain walls and stone. The glass curtain walls, holding an important position in the elevation pattern, apply the combination of modern building materials and conventional patterns to presenting both the spirit of times and the national specialty. Taking the national conditions and cost into consideration, we designed a unique window-type curtain wall system. The system decomposes the curtain wall into 4m×3.6m units, fixed on the steel frame by the conventional method to make the construction easier and save the cost. The central lobby uses the dot-connected glass curtain walls, which gives a good effect.

望京科技园二期

北京

Wangjing Science & Technology Park (Phase II), Beijing

2004

望京科技园二期位于望京新兴产业区北部，五环路南侧，是一个配套齐全的办公建筑。该建筑地处城市边缘，容积率较低，环境较好。

本工程由三幢平面类似的建筑和一个连接体组成。其中两幢以直线形式排列在用地北端，另一幢位于用地南侧，中间是一个类楼梯的连接体。这个连接体内是门厅和咖啡厅、展厅及管理用房。由于消防的要求在北侧有一个消防通道，于是 A 栋的主入口在二层，连接体内部提供了一个从一层至二层的类楼梯空间，既满足了使用要求，也丰富了建筑空间。在建筑形体上，两个主体量分别体现了矩形沿折线轨迹翻转的感觉，另一个体量，是由密肋玻璃与全透明幕墙组成的从半透明到全透明的体量。主体块外墙采用低辐射中空印刷玻璃，光线折射的结果和印刷图案共同作用造成一个种独特的效果。大悬挑部分强化了入口和重点部位的体型效果。

本工程地下 1 层、地上 6 层。地下一层为车库和机房及职工餐厅，地上除 A 栋一层为会议中心，B 栋六层除健身中心外均为办公楼。

该工程占地面积 25916 平方米；建筑面积 46297 平方米，其中地上 37048 平方米，地下 9248 平方米。

Wangjing Science & Technology Park (Phase II) is on the south side of the 5th Ring Road, in the north of Wangjing Emerging Industry Zone and is an office building with complete facilities. Located on the edge of the city, the building has a relatively low FAR and a better environment.

The project consists of three buildings with similar plans and one connection structure. Two buildings are arranged in a straight line in the north of the lot and the other one is in the south, between which is a stair-type connection structure. Inside the structure are an entrance hall,a cafe, an exhibition hall and an administrative center. Since there is one fire passage required by the fire safety, the main entrance of Building A is on the second floor, while the structure provides the stair-type space from the first floor to the second floor to meet the requirement of use as well as enrich the architectural space. In terms of architectural form and structure, the two main masses represent respectively the feeling of a rectangle overturning along the broken-line track, while the other one is a mass translating from semi-transparent and to fully-transparent, composed of close-ribbed glass and full-transparent curtain wall. The main block is externally finished with Low-E insulating printed glass to create a unique effect to the combination of light reflex and printed pattern. The large cantilever part enhances the size effect of the entrance and key locations.

The project includes one level of basement and six floors. The basement is used for garage, machinery room and dining hall for employees, while all the floors are used for offices, except for 1st floor of Building A, as the conference center, and 6th floor of Building B, as the fitness center.The project has a land area of 25,916m^2 and a floorage of 46,297m^2, including 37,048m^2 for the superstructure and 9,248m^2 for the basement.

外景 / Exterior view

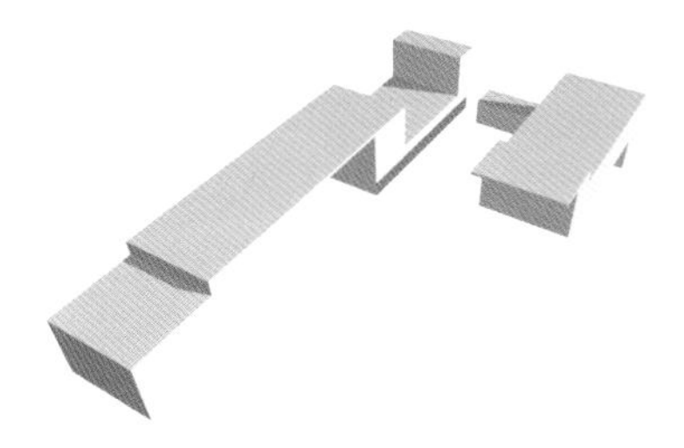

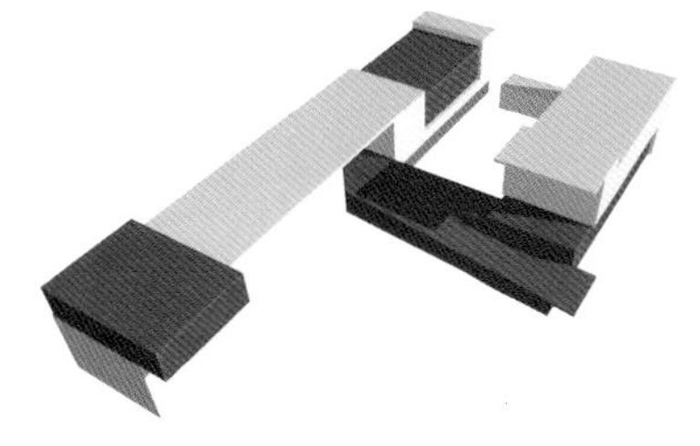

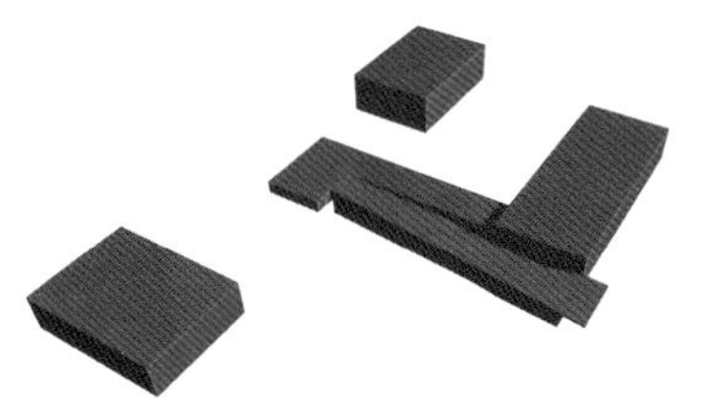

分析图 / Digram
外景 / Exterior view

总平面图 / Master plan
B、C 座西立面图 / West elevation of block B & C
A、B 座北立面图 / North elevation of block A & B
内景 / Interior view

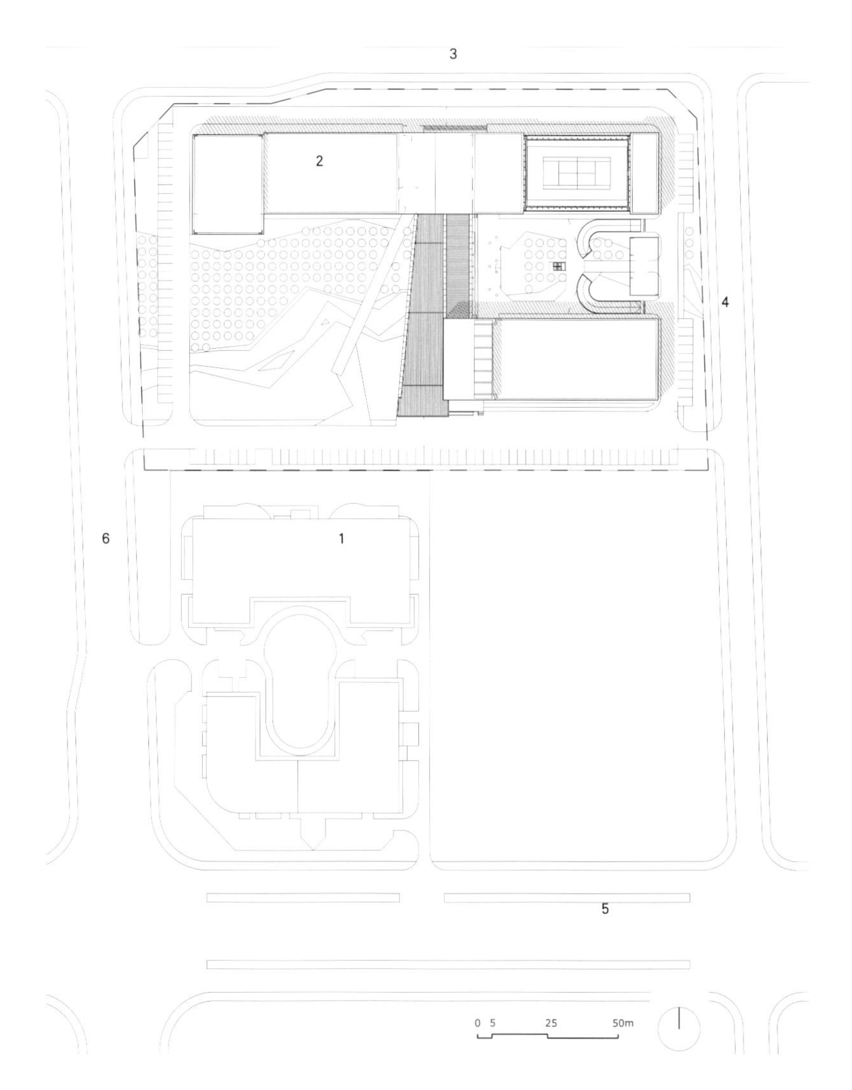

1 望京科技园一期 / Wangjing Science & Technology Park (Phase I)
2 望京科技园二期 / Wangjing Science & Technology Park (Phase II)
3 北五环南侧辅路 / South Side Road of the North 5th Ring Road
4 朝科内部 D 路 / Chaoke Inner D Road
5 望京外环路 / Wangjing Outer Ring Road
6 科技开发区 2 号线 / Line 2 of Science & Technology Development Zone

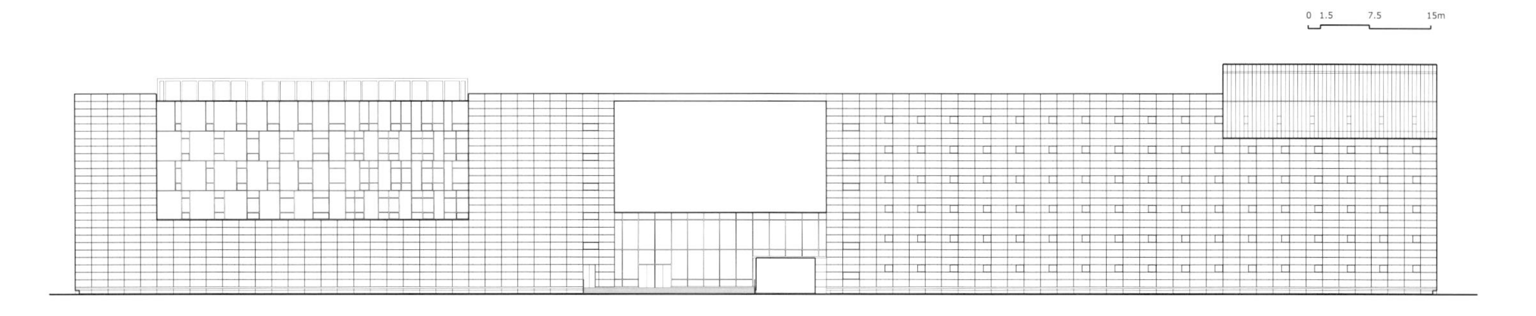

外景 / Exterior view
首层平面 / The 1st floor plan
标准层平面图 / Plan for standard floor

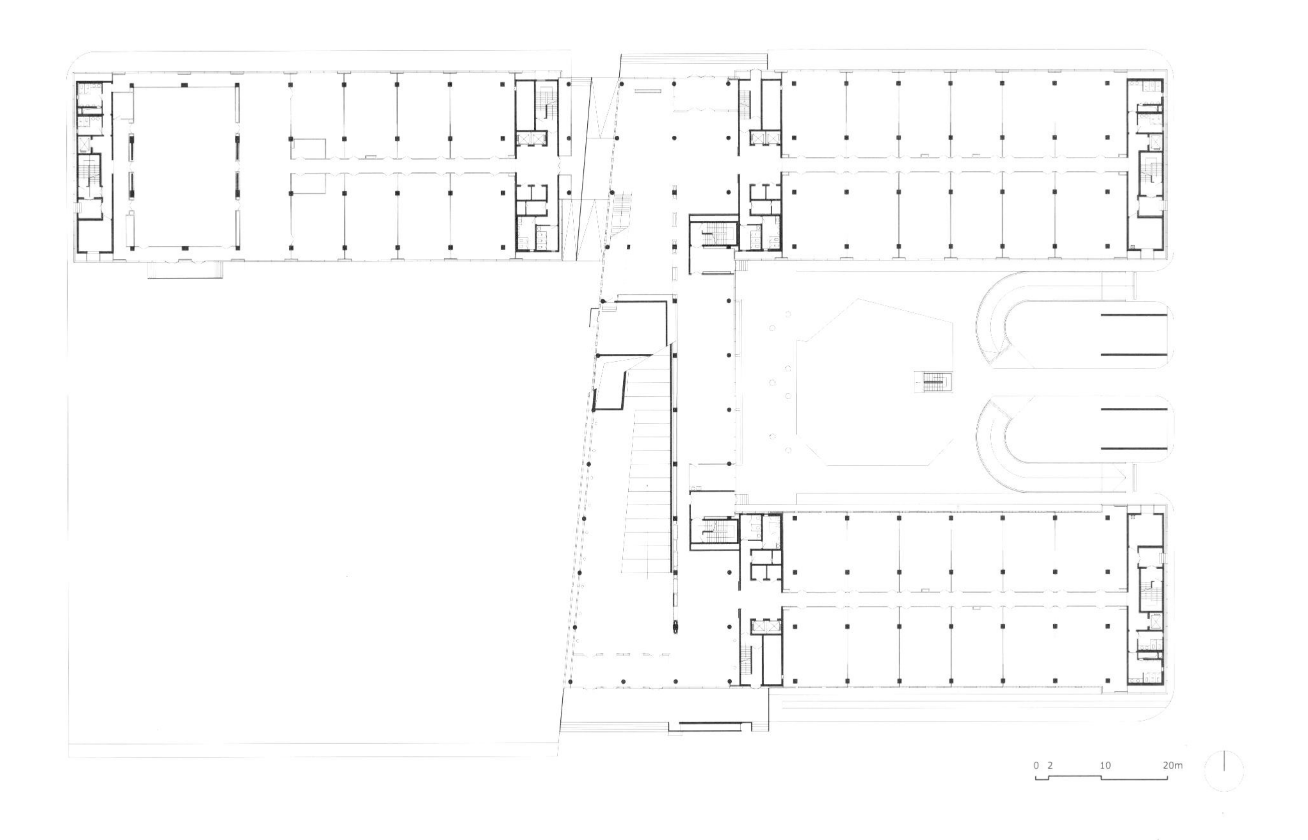

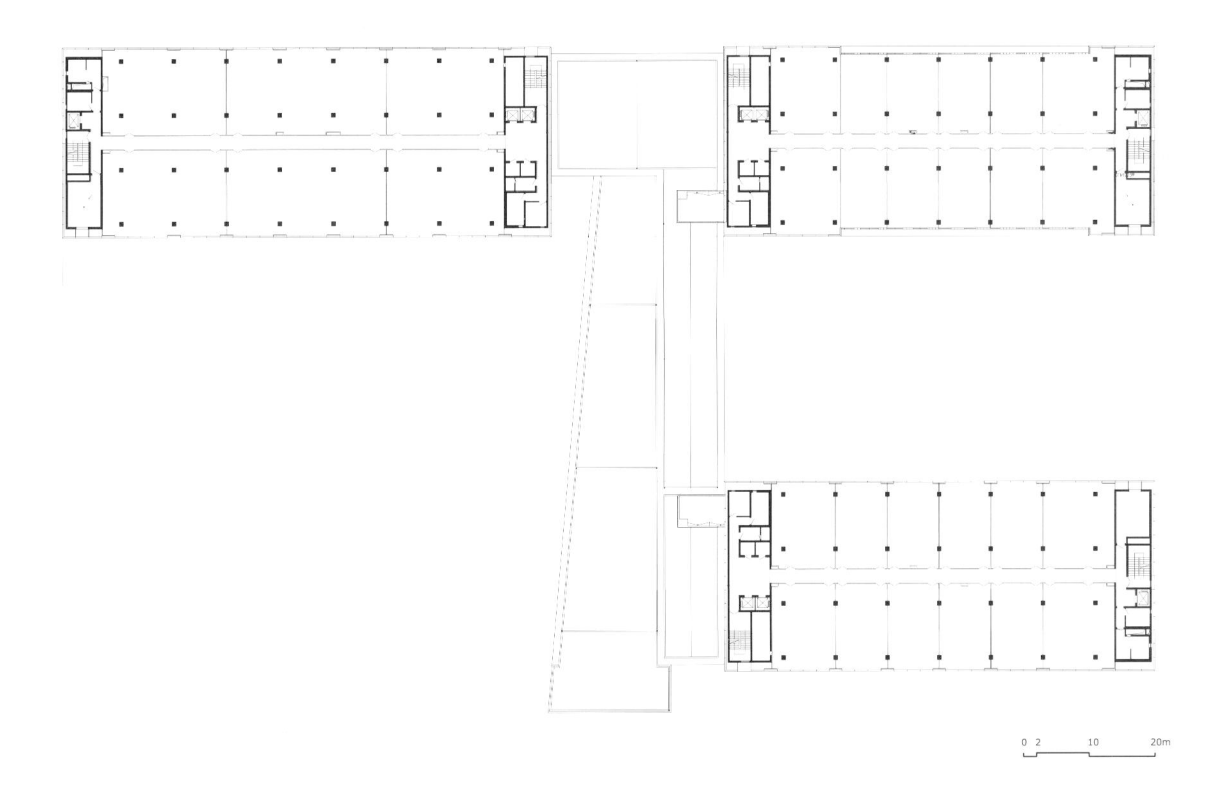

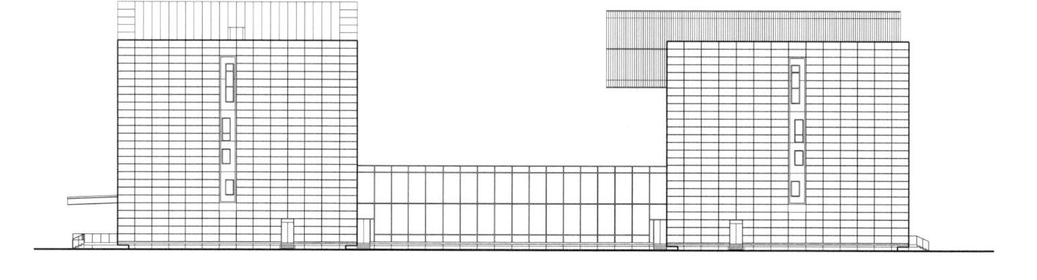

外景 / Exterior view
B、C 座东立面图 / East elevation of block B & C

外景 / Exterior view

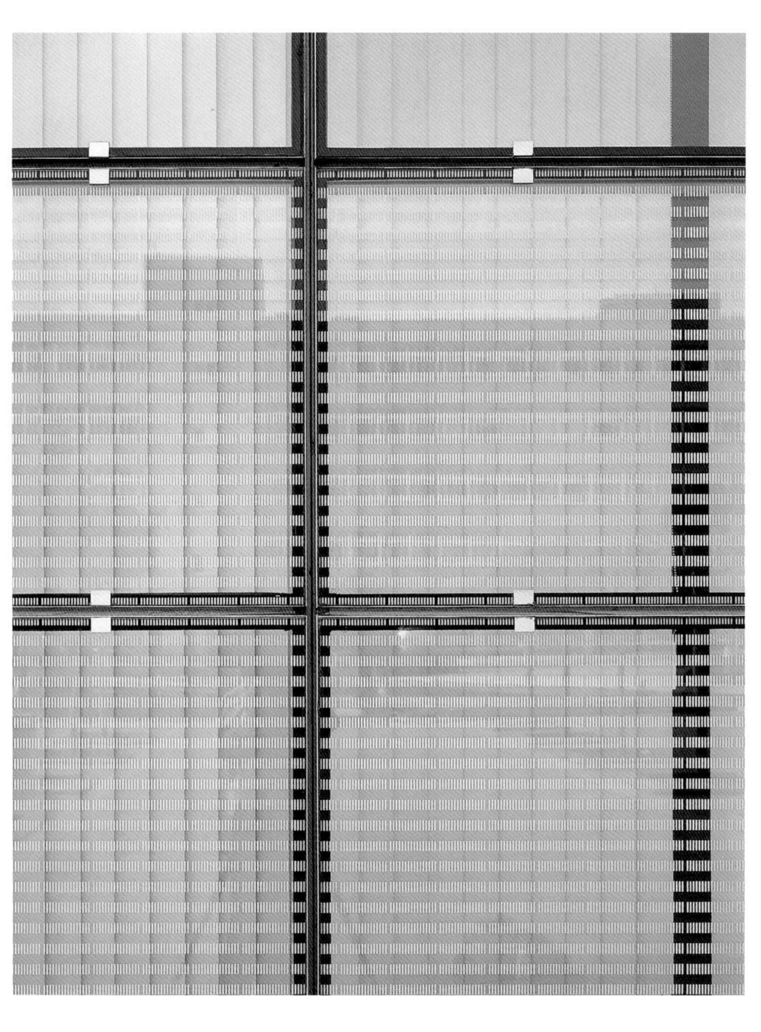

细部 / Details
外景 / Exterior view

京A 08422

外景 / Exterior view

外景 / Exterior view
内景 / Interior view
剖面 / Section

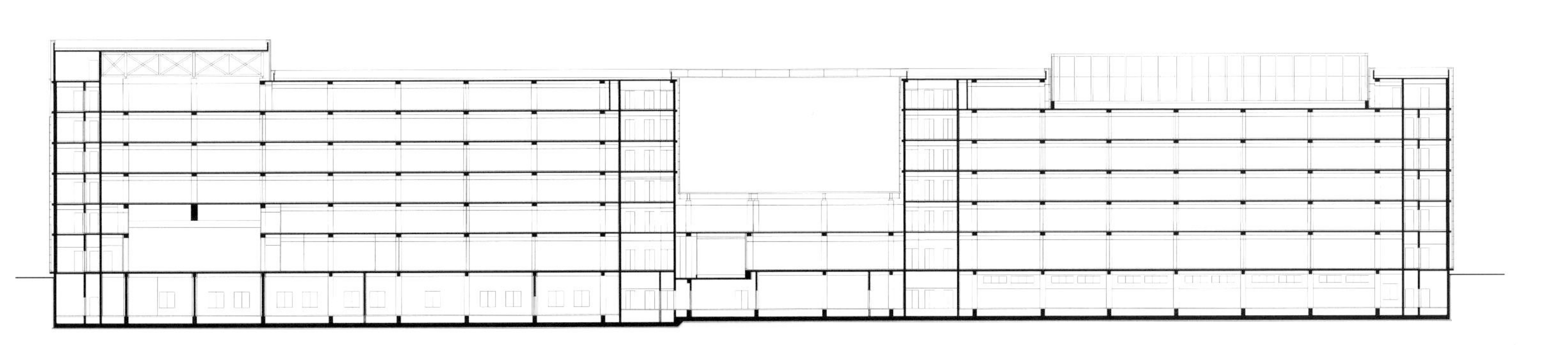

外景 / Exterior view

五棵松棒球场　　北京

Wukesong Baseball Field, Beijing

2007

五棵松棒球场是第 29 届奥运会棒球比赛的主赛场，为临时建筑，奥运会之后拆除，还原为供百姓日常进行文体活动的区域。五棵松棒球场共有 15000 个座席，在国际棒联的强烈要求下将其分为 12000 座（1 号）和 3000 座（2 号）两个棒球场，与训练场一起三个场地由南向北线性排布，全部采用钢结构建造。1 号、2 号场地包含运动员、裁判、赛事管理、VIP、媒体及安保等各类奥运比赛所需的功能用房，关系复杂。鉴于奥运会的比赛要求，棒球场虽为临时建筑，但其比赛场地及比赛设施是世界一流的，专业的草和红土花费了大量的财力，每个场地周围配备的 6 根高杆投光灯保证了比赛的照度需求，但它的临时性决定了建筑必须是节俭的，尽可能采用成熟技术、简单工艺及经济材料。布局上平层分流，观众和特殊人员的流线交叉通过运营管理手段解决。不设电梯，残疾人通过残疾人坡道式轮椅爬楼车到达指定位置。结构设计上选择钢结构，保证了拆卸的便捷及材料的持续性，12000 座棒球场 70% 看台租用专业脚手架临时搭设，配备低靠背硬塑座椅。计时计分显示屏及颁奖台升降旗等大量运动工艺设施赛前临时租用，大大降低了造价。

五棵松文化体育中心内的重心是篮球馆，因此棒球场采用“隐”的设计手法，在深灰色建筑外墙外局部罩一层疏密相间的金属网，通过绿色的变化肌理形成迷彩外衣，作为绿色藤蔓植物的依附物，这样棒球场在夏季奥运会来临之际将会融入郁郁葱葱的绿色环境之中，达到和谐自然的目的。

外景 / Exterior view

Wukesong Baseball Field was the main contest arena for the baseball of the 29th Olympic Games. As a temporary facility, it was removed and restored for the public to use as a recreation area. Wukesong Baseball Field had in total 15,000 seats, which was separated, as strongly required by the International Baseball Federation, into two base fields (1#) 12,000 seats and (2#) 3,000 seats which were arranged, together with the training field, in linear from south to north, and built with the steel structure. 1# and 2# arenas included the functional rooms for athletes, referees, contest management, VIP, media, and other purposes for the Olympic contents, showing a complex relationship. Due to the contest requirement of the Olympic Games, the baseball field's contest arena and facilities were world-class, with huge financial investment, which were made on the professional lawn and red soil. Each arena was surrounded with 6 high-rise posts of spot lights to ensure the lighting requirement of contest. However, as temporary facilities, the architecture had to be economical and thus efforts were made to use mature technologies, simple processes and economic materials. In terms of layout, horizontal distribution and intersection for flow lines of audiences and specific staff were resolved by means of operation management. Since no elevators were set up, the disabled had to reach the designated position through the ramp-type wheel-chair stair lifts. The steel structure was designed to ensure the easy disassembling and recycling of materials. For the 12,000-seat baseball field, 70% of the bleachers were built by leasing the professional scaffolds and equipped with low-back hard plastic seats. Timing and scoring display screens, awarding platform and other sports facilities were leased before contest so as to much cost.

总平面图 / Master plan
外景 / Exterior view

0 25 125 250m

Since the focus of Wukesong Culture and Sports Center is the basketball gymnasium, the “concealing” design method was used for the baseball field: the dark-grey external wall of the building was partially covered with one layer of metal net of different densities; the changing tissue with dark grey in green formed a pattern-painting coat, as the dependent frame for green vines. Thus, during the Summer Olympic Games, the baseball field was integrated into a green environment for the harmonious and natural purpose.

Padres Club House
教士队更衣室
A8-1
A8-2
A8-3
East Stand Entry

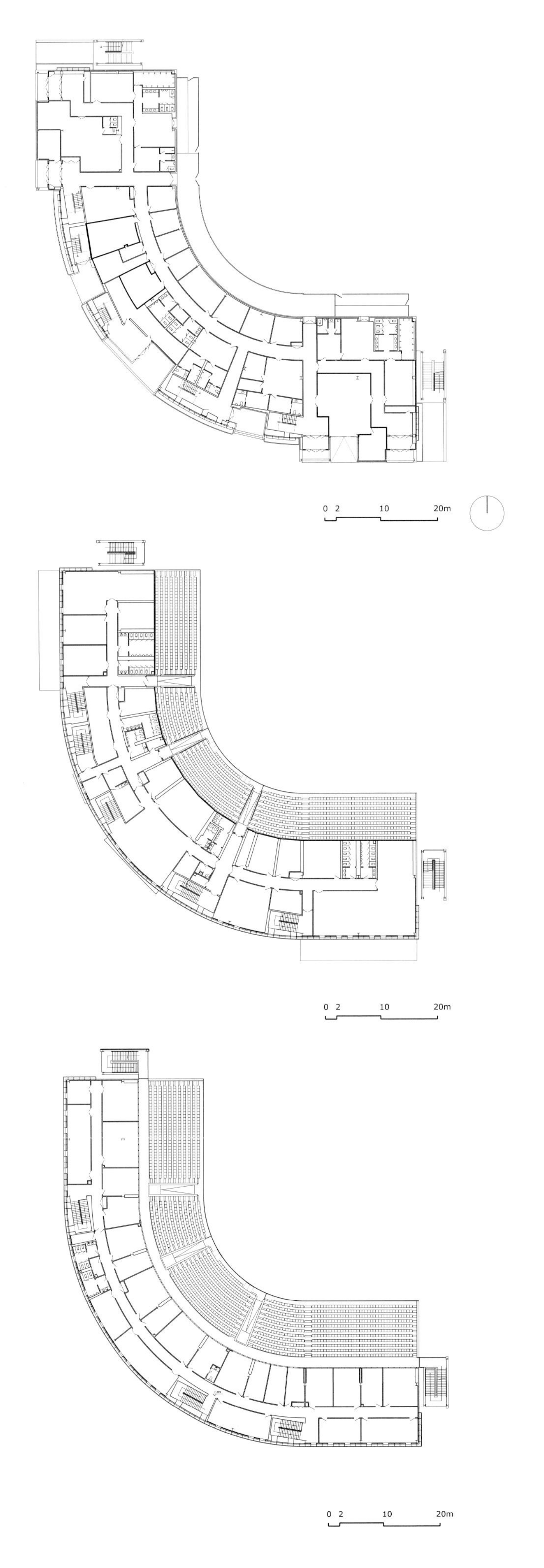
0 2 10 20m
0 2 10 20m
0 2 10 20m

12000 座棒球场首层平面图 / Plan for 1st floor of 12000-seat baseball field
12000 座棒球场二层平面图 / Plan for 2nd floor of 12000-seat baseball field
12000 座棒球场三层平面图 / Plan for 3rd floor of 12000-seat baseball field
普通观众看台剖面 / Auditorium section
贵宾看台剖面 / VIP auditorium section
外景 / Exterior view

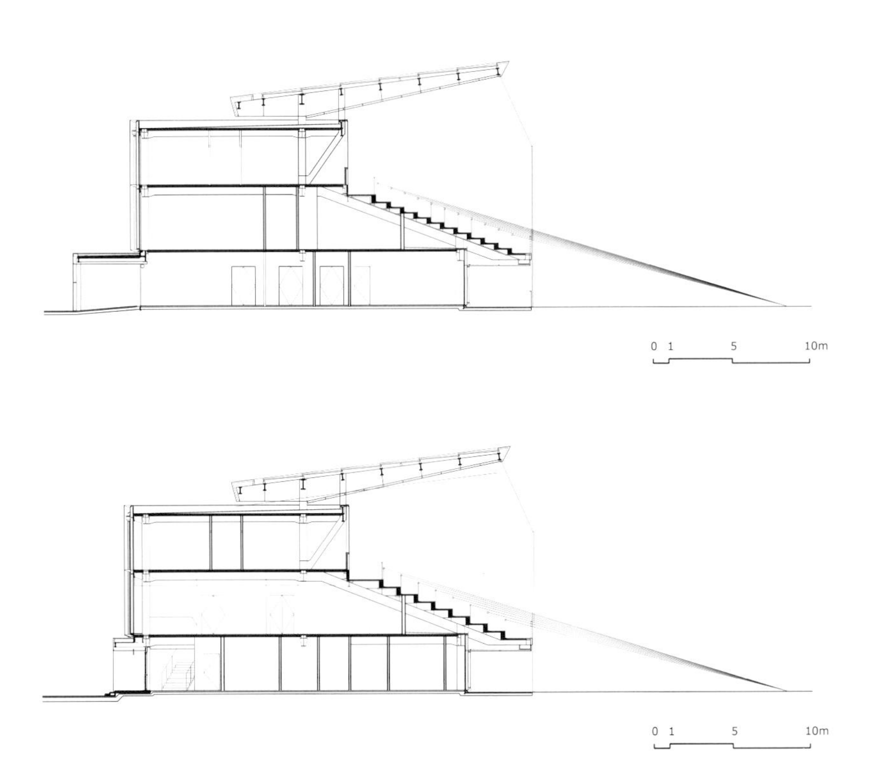

外景 / Exterior view

(12000 座) 棒球场外层西立面图 / West elevation of the exterior for the 12000-seat baseball field

(12000 座) 棒球场基层西立面图 / West elevation of the base for the 12000-seat baseball field

(12000 座) 棒球场外层南立面图 / South elevation of the exterior for the 12000-seat baseball field

(12000 座) 棒球场基层南立面图 / South elevation of the base for the 12000-seat baseball field

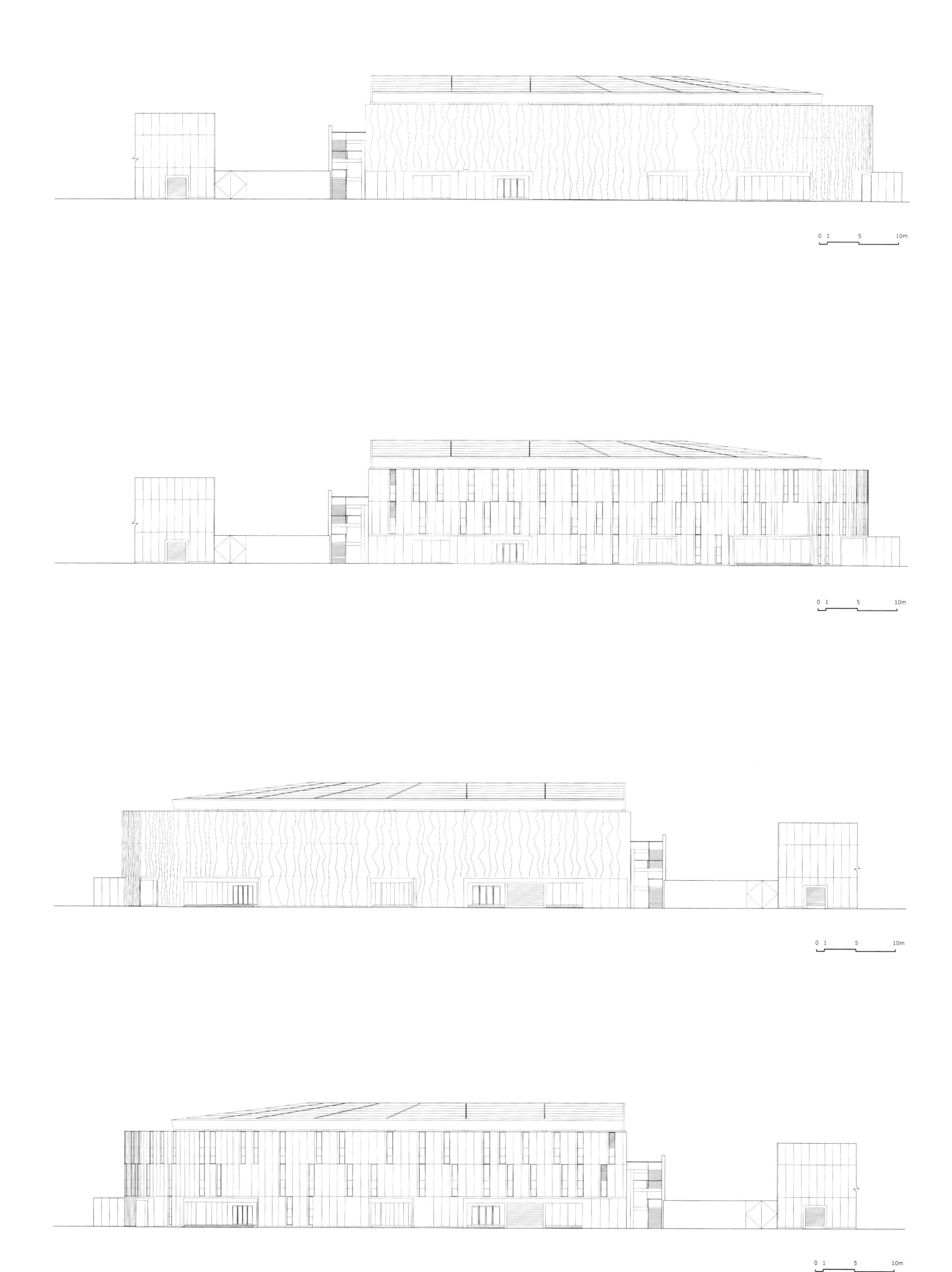
0 1 5 10m
0 1 5 10m
0 1 5 10m
0 1 5 10m

外景 / Exterior view
(12000 座）棒球场外立面基层展开图 / Base spreading of exterior elevation for the 12000-seat baseball field
(12000 座）棒球场外立面面层展开图 / Facade spreading of exterior elevation for the 12000-seat baseball field
(12000 座）棒球场东立面图 / East elevation of the 12000-seat baseball field
(12000 座）棒球场全场南立面 / South elevation of the 12000-seat baseball field

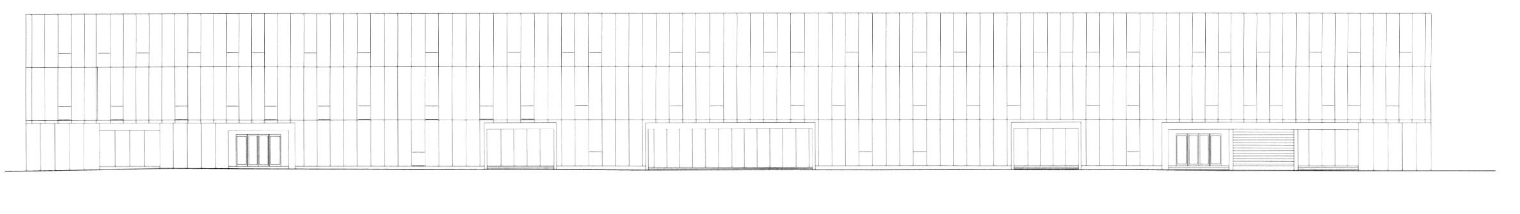
0 1 5 10m

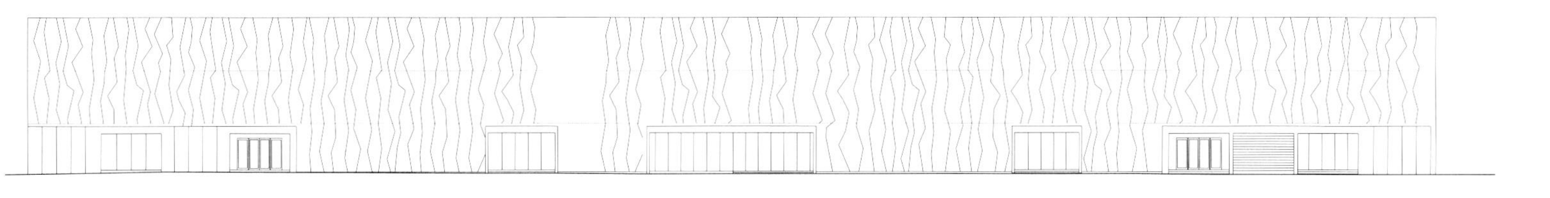
0 1 5 10m

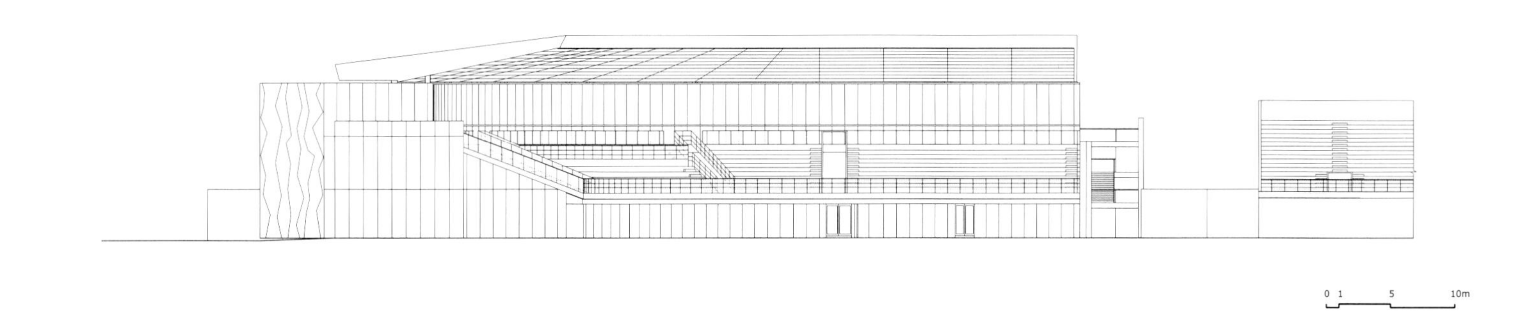
0 1 5 10m

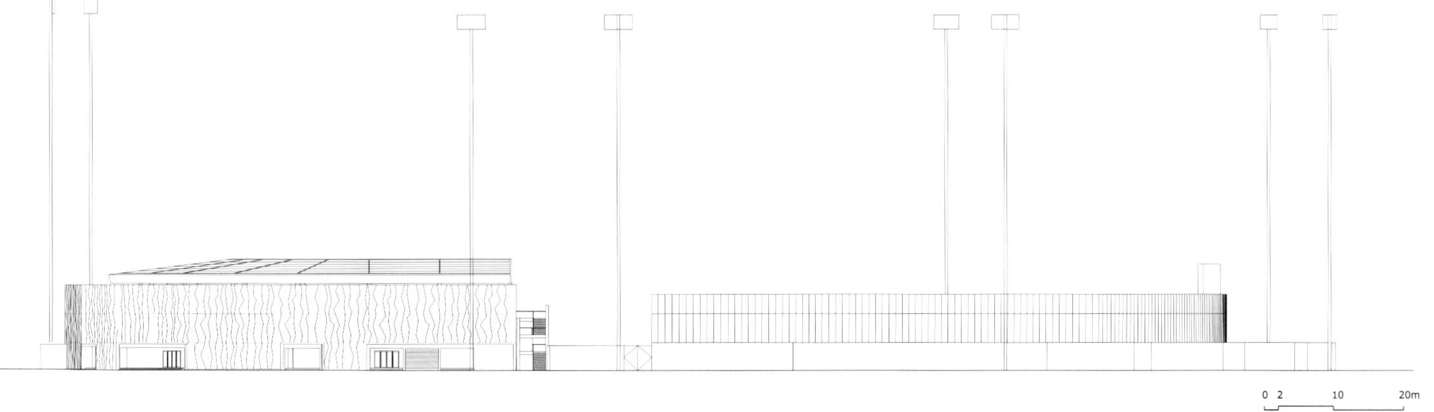
0 2 10 20m

外景 / Exterior view

局部立面 / Partial elevation
外景 / Exterior view

上海青浦体育馆、训练馆改造

上海

Shanghai Qingpu Gymnasium and Training Center Renovation, Shanghai

2008

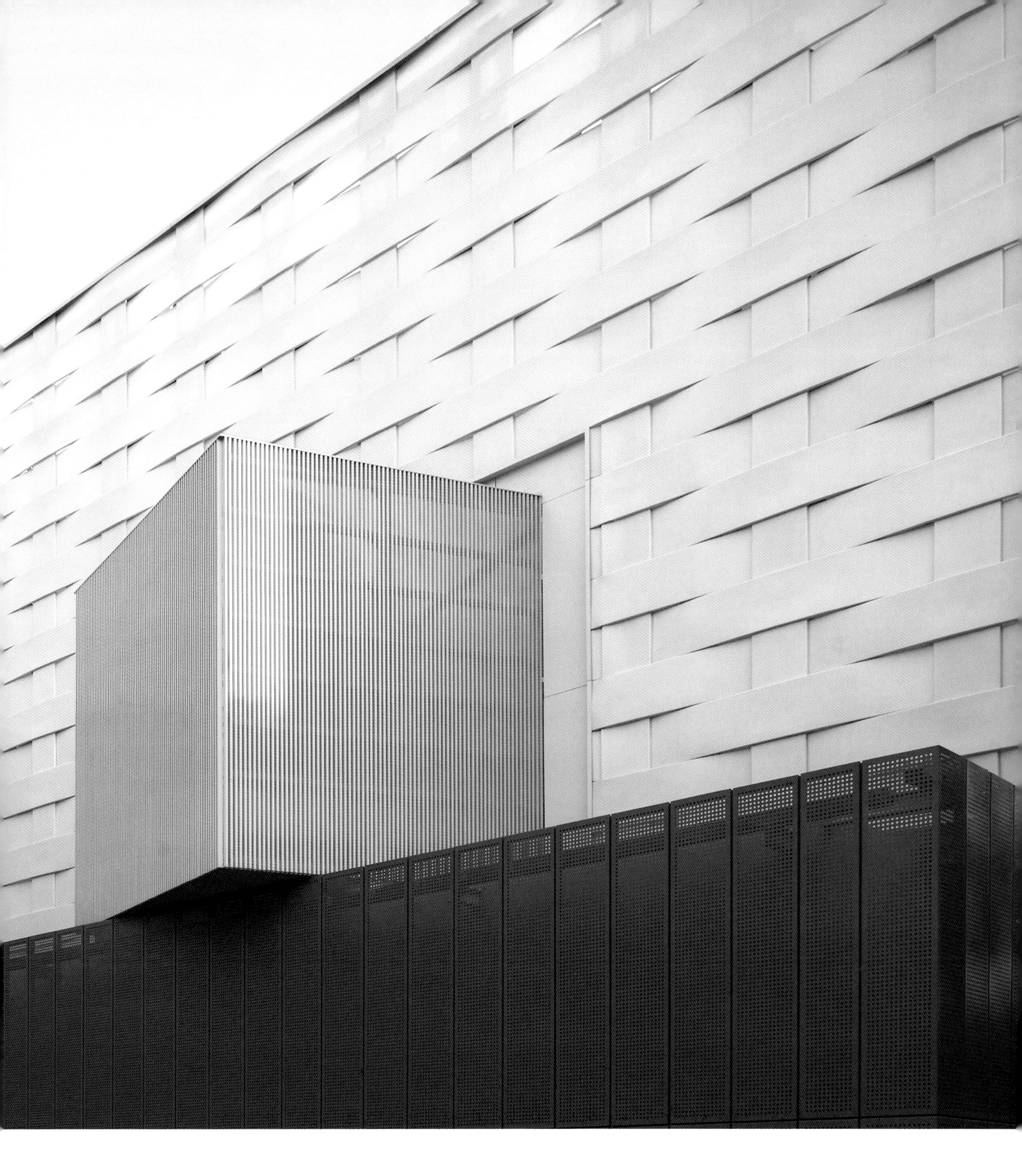

上海青浦体育馆、训练馆位于上海青浦区的旧城内，两条城市道路交叉路口的东北侧。建筑面积 8100 平方米。

两个馆均建于 1980 年代早期。由于年代久远，原有建筑、设施破旧，建筑立面造型存在较大缺陷，不能满足快速发展的城市的要求。政府希望通过这次改造彻底改变原建筑的面貌，同时改善其内部设施以便为市民提供一个运动健身的场所。

本项目设计克服了原始设计资料缺失、造价低廉等不利条件，合理巧妙地应用新材料、新技术以及新构造做法，采用独创的聚碳酸酯板编织外墙，不仅保证了建筑内部的自然采光效果，而且创造了独具一格的建筑形象。同时本项目利用金属格栅和穿孔铝板将原建筑造型上存在缺陷的室外楼梯、入口雨罩、空调室外机和旧墙包裹起来，使建筑在保留原有外墙和建筑构件的同时焕然一新。

本工程建成后，使这一地区的城市面貌发生了很大变化，同时它已经成为上海市青浦区一处深受市民欢迎的体育健身场所。

Shanghai Qingpu Gymnasium and Training Center are located inside the old city area of Shanghai Qingpu District, on the northeast side of the intersection between two city roads, with a floorage of 8100m^2. The two facilities were built in the early 1980s. Being built long time ago, the existing buildings and facilities are outdated, with major defects observed on the elevation modeling, and have failed to meet the requirements of the city in rapid development. The government expected to change thoroughly the appearance of the existing buildings through the renovation and improvement of the internal facilities for the purpose of providing the citizens with an arena for sports and recreation. By overcoming such unfavorable factors as lack of original design data and low cost, the renovation project was designed rationally and cleverly with new materials, new techniques and new building methods. The unique exterior wall woven with polycarbonate sheets not only ensured the effect of natural lighting inside the building, but also created a unique architectural image. Additionally, with the metal grills and perforated aluminum panels, the projects covered up the exterior stair, entrance shelter and outer A/C units of the original building, which were defected in the old modeling, for a new appearance of the building while maintaining the existing exterior wall and structural components. Upon completion of the projects, a great change was observed in the urban appearance of the area; it also became an arena favored by the citizens for sports and fitness-keeping in Qingpu District, Shanghai.

外景 / Exterior view

改造前外景 / Exterior view before renovation
改造后外景 / Exterior view after renovation

改造前内景 / Interior view before renovation

改造后内景 / Interior view after renovation

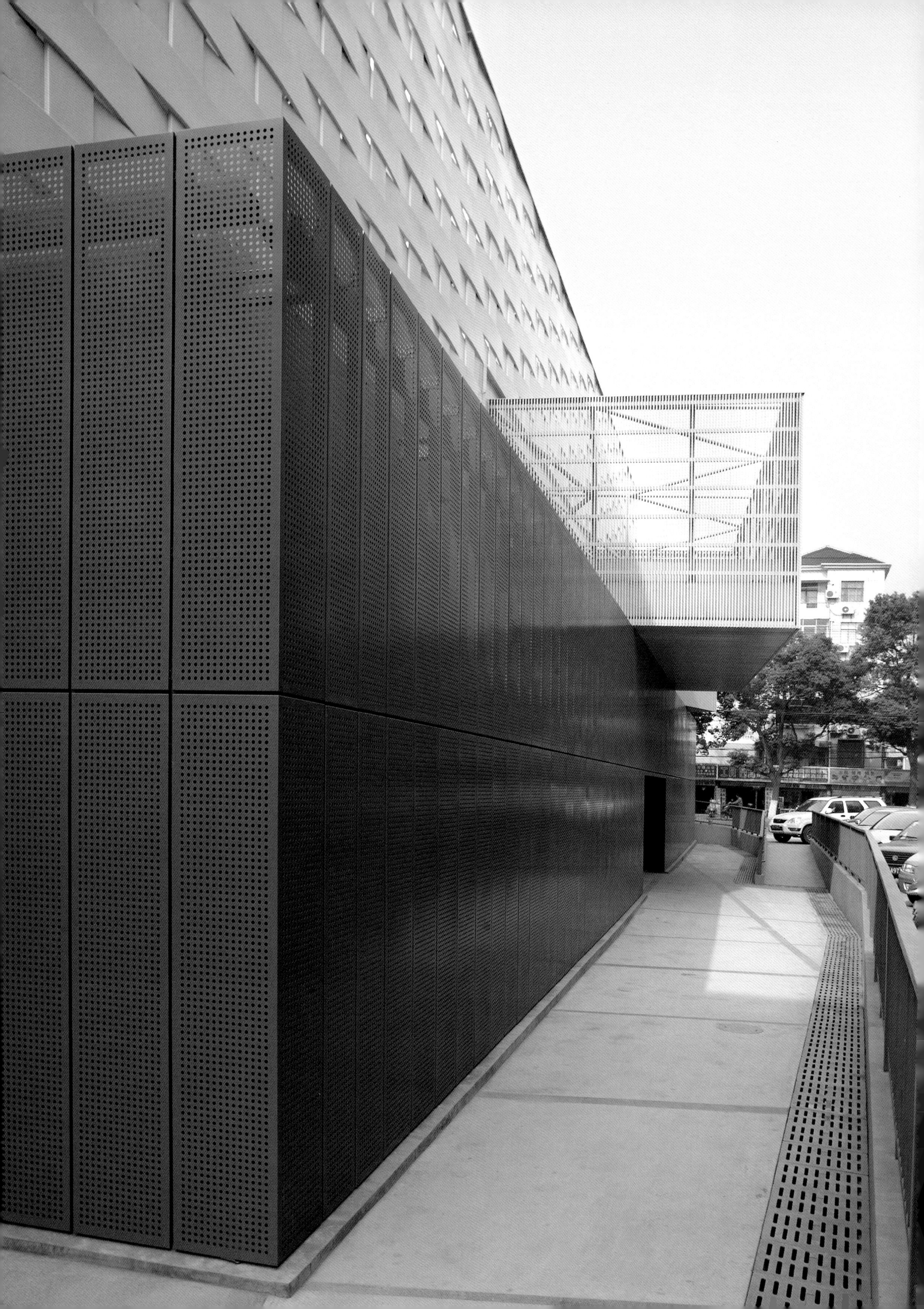

总平面图 / Master plan
训练馆首层平面 / Plan for the 1st floor of the training center

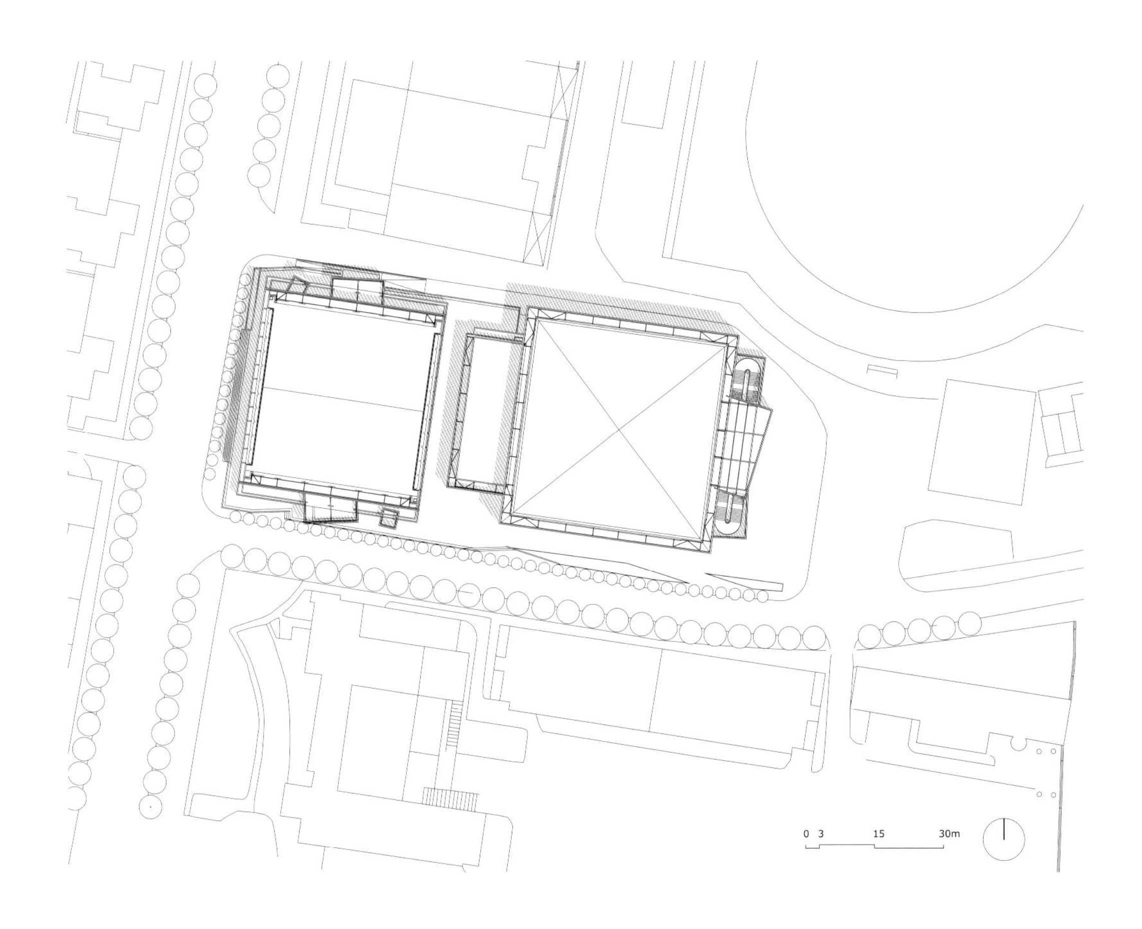

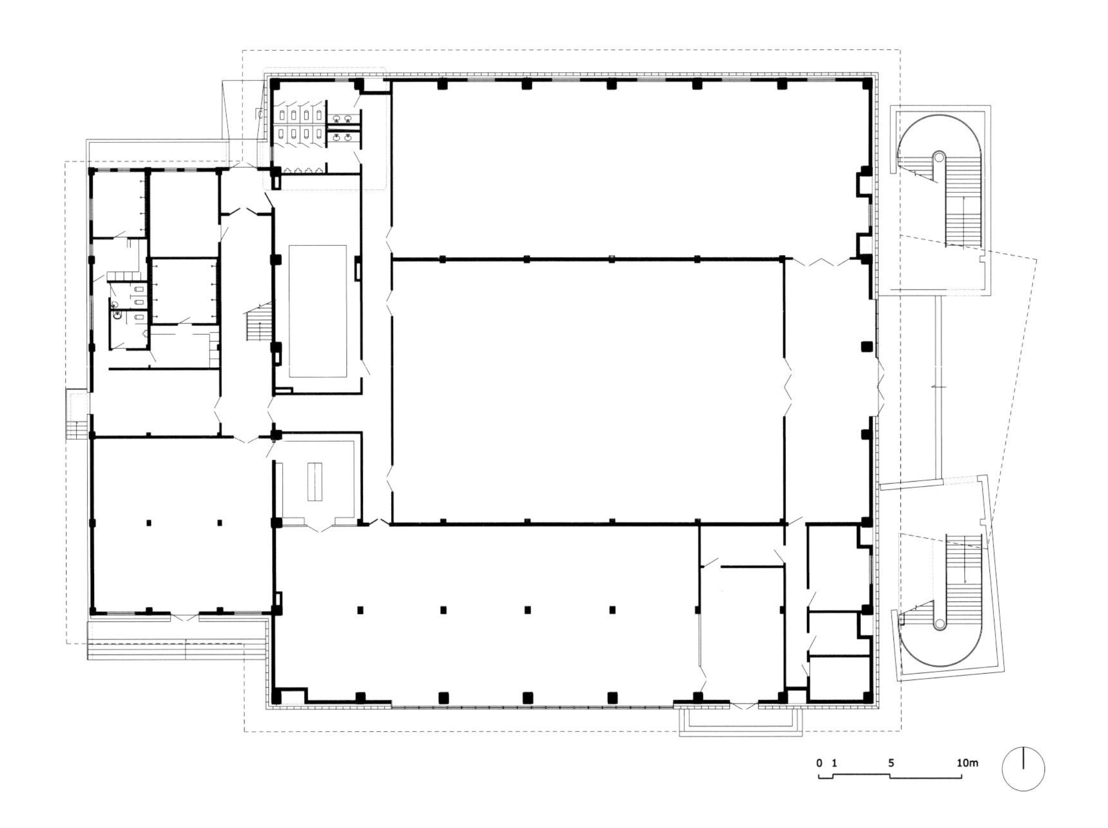

训练馆东立面图 / East elevation of the training center
训练馆南立面图 / South elevation of the training center
分析图 / Diagram

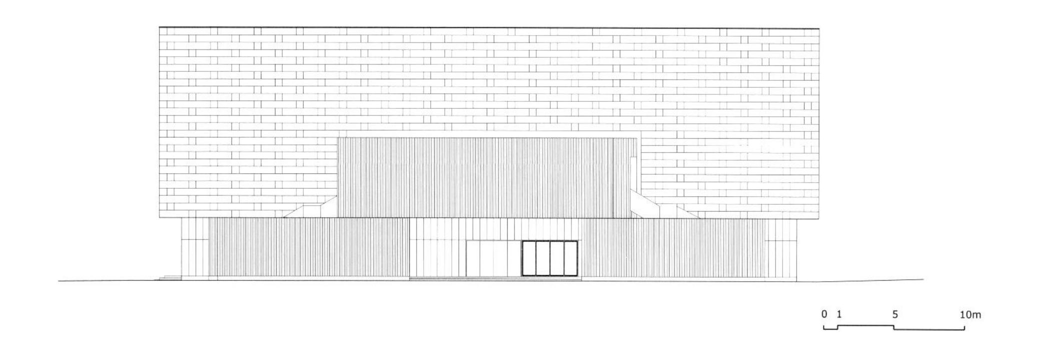

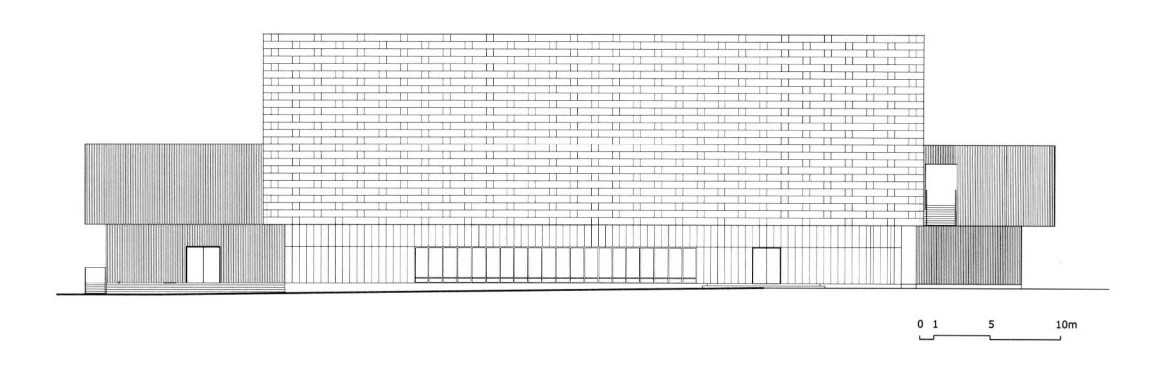

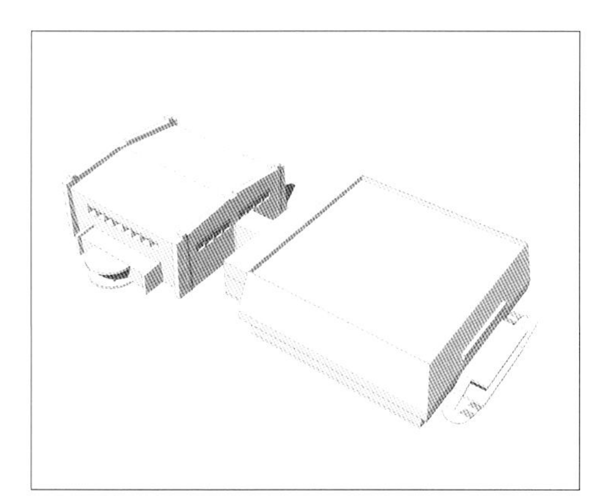

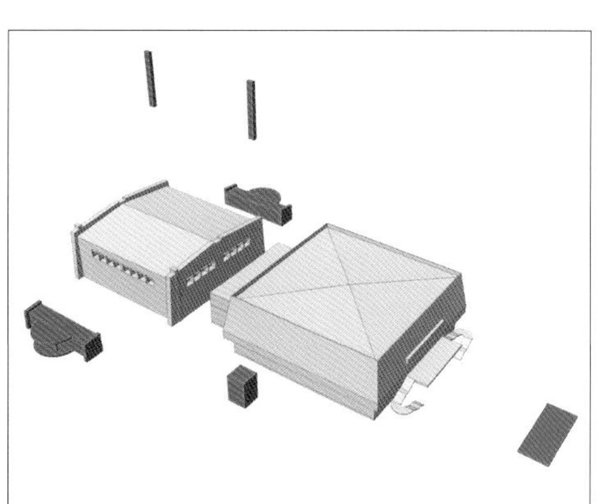

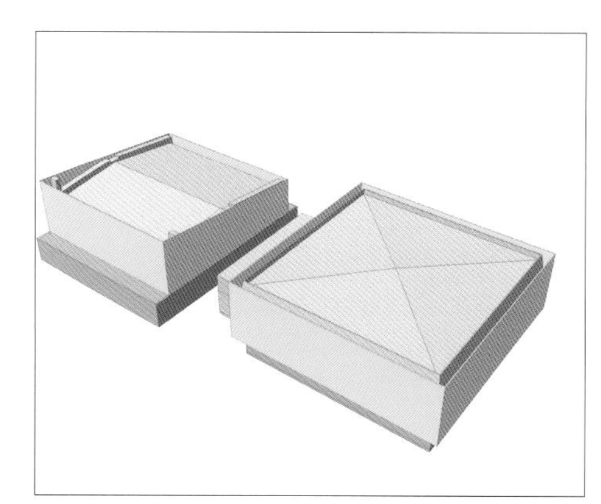

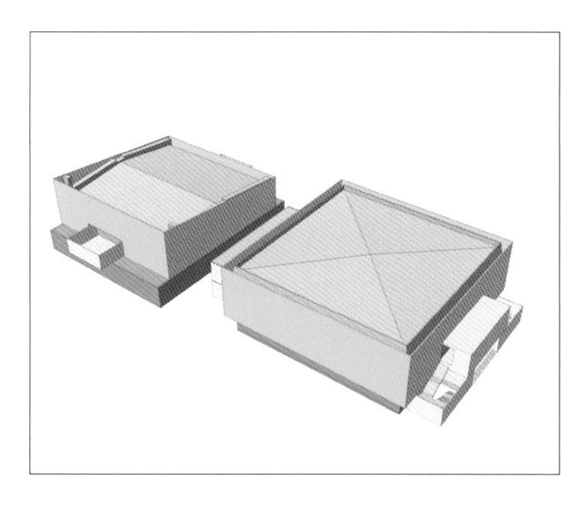

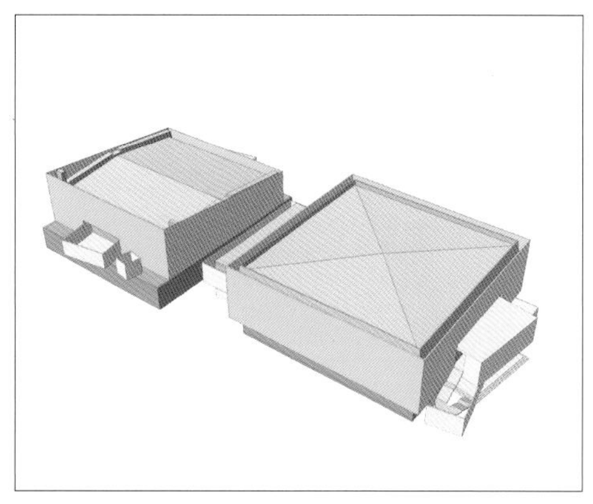

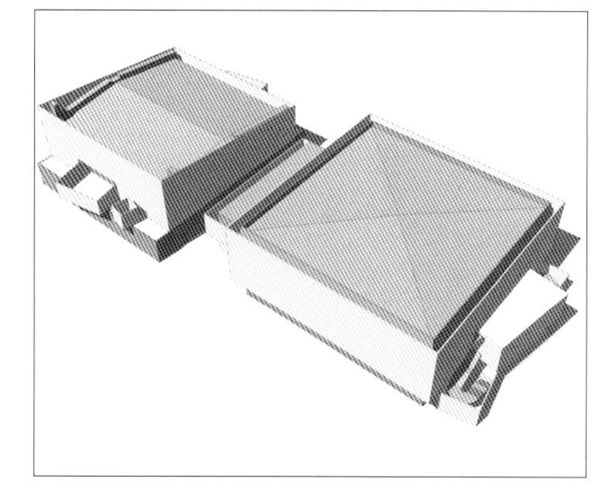

体育馆一层平面 - 活动座椅 / The 1st floor plan of gymnasium-removeable seat
体育馆看台平面 - 活动座椅 / The stand floor plan of gymnasium-removeable seat

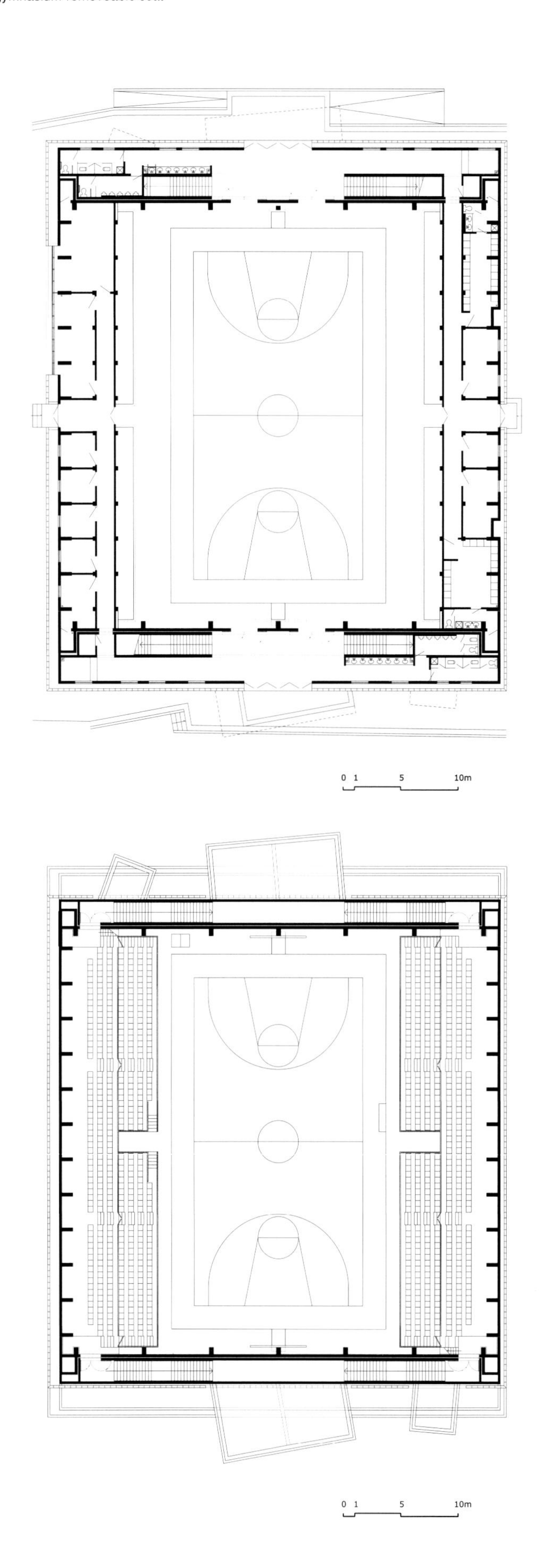

体育馆南立面图 / South elevation of Gymnasium
体育馆西立面图 / West elevation of Gymnasium
外景 / Exterior view

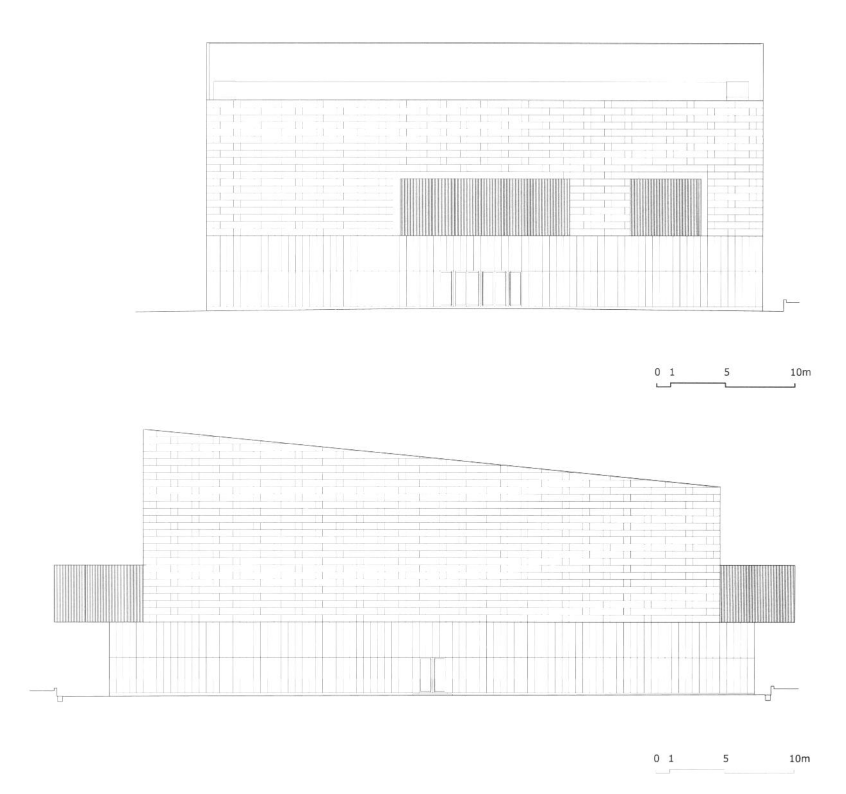

改造前内景 / Interior view before renovation
改造后内景 / Interior view after renovation
体育馆剖面图 / Section

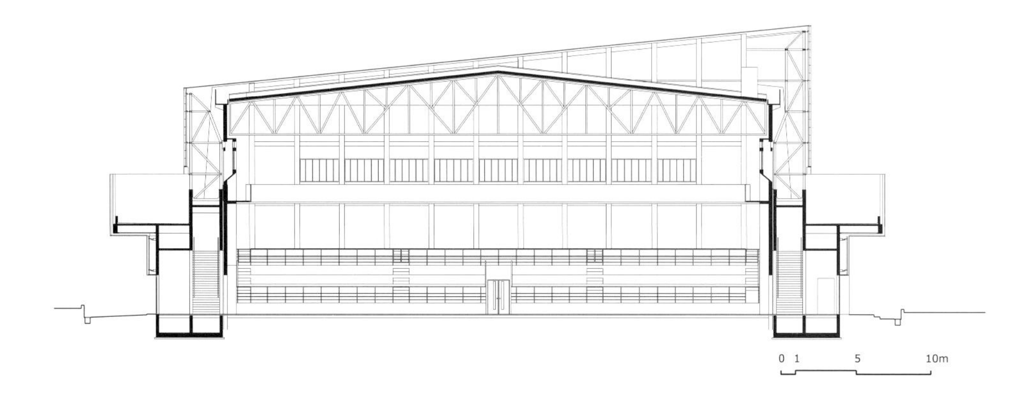

细部 / Details
外景 / Exterior view

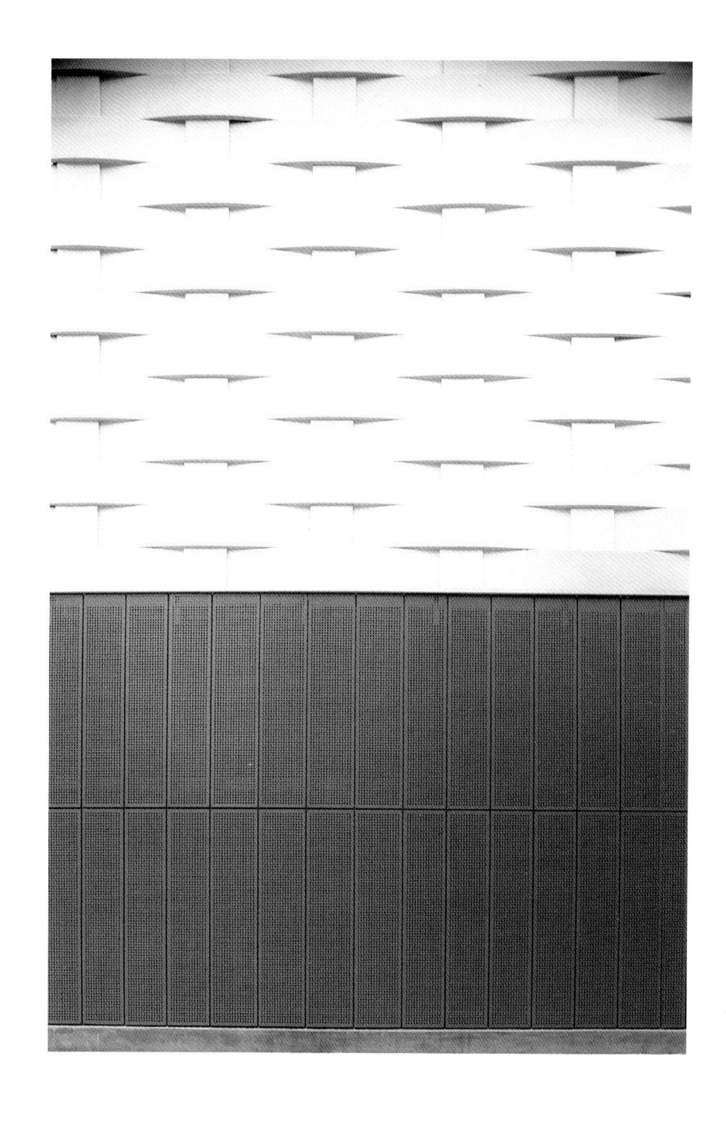

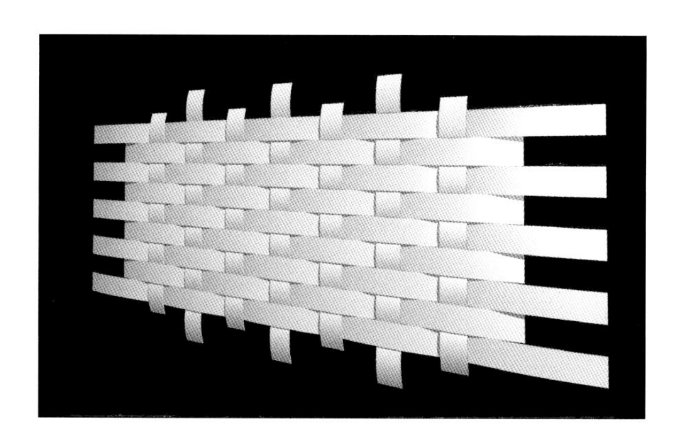

外景 / Exterior view
体育馆剖面图 / Section

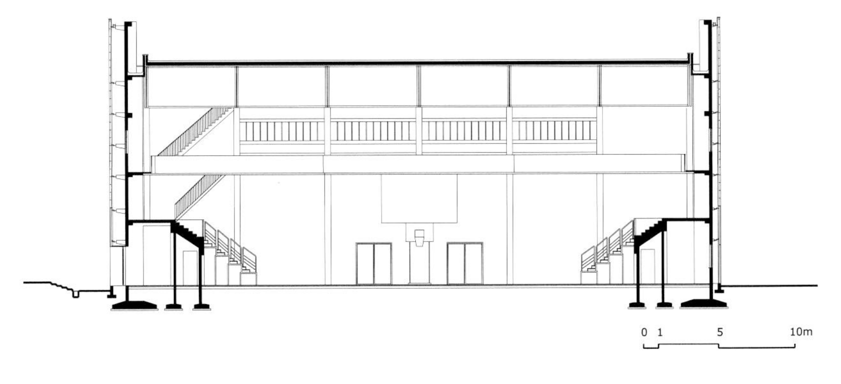

外景 / Exterior view

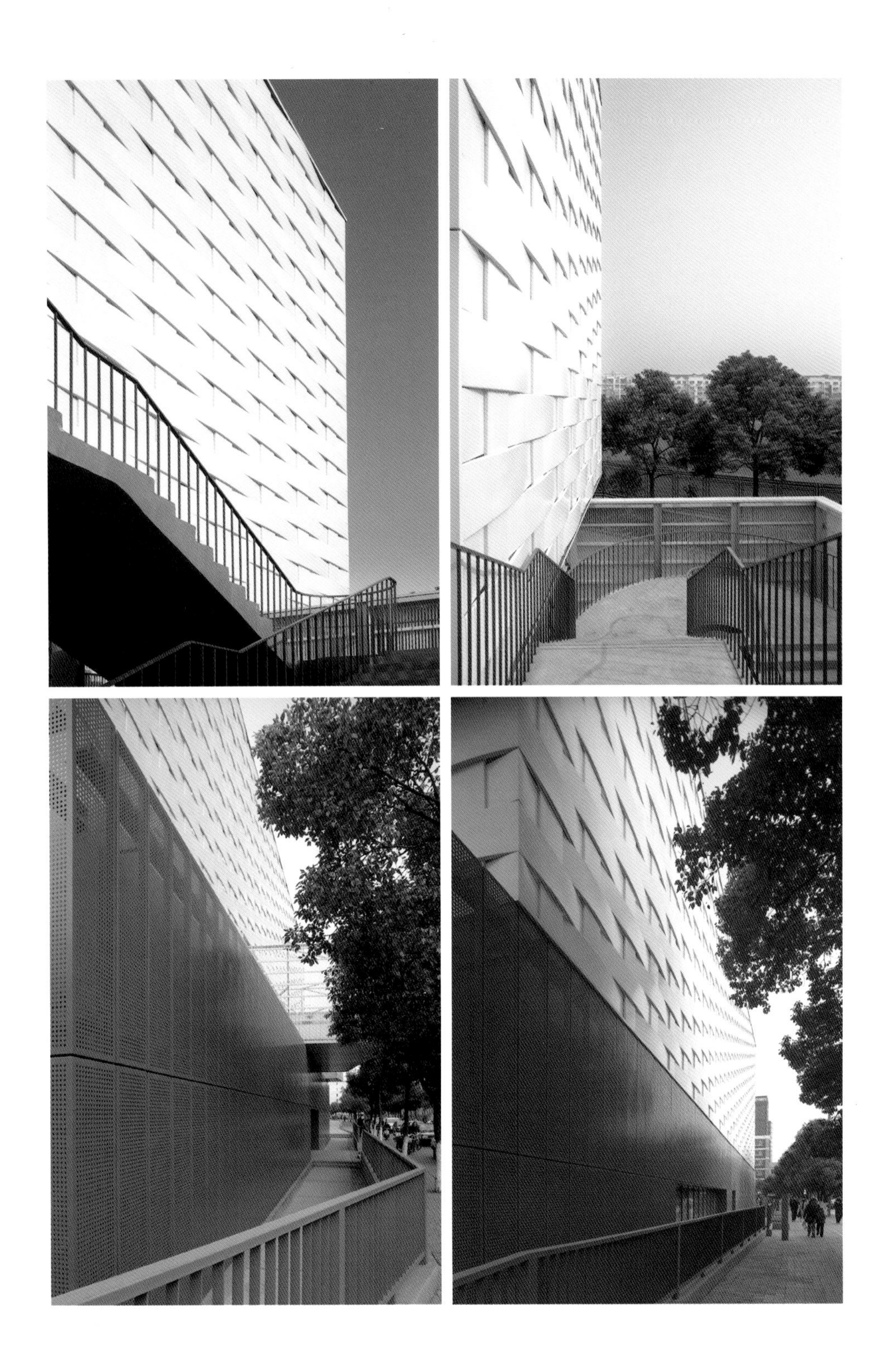

外景 / Exterior view

上海 2010 世博会 UBPA 办公楼改造　上海

Shanghai Expo 2010 UBPA Office Building Renovation, Shanghai

2010

2010 年上海世博会城市最佳实践区北部街区 UBPA 办公楼（以下简称 UBPA 办公楼）是一个旧建筑（理化测试实验中心）改造工程。现状建筑是一个普通多层建筑，外观陈旧、无特色。除了首层层高为 4.2 米外，二至五层层高为 3.1 米，六、七层层高为 3.0 米，作为办公建筑层高较低。同时，整个建筑公共设施标准较低，如电梯、卫生间等设施已经不能满足现代办公的要求。现状建筑东北侧突出在红线外，为减少对道路用地的占用，业主要求改造时将首层做成骑楼形式。

通过改造该建筑将成为城市最佳实践区一永久办公建筑。

建筑体形基本沿袭旧有建筑的体型关系，为一个平面“U”形的多层板式建筑。立面处理遵循简洁、经济但不失美观的原则，各个方向立面均采用在墙面上均质开窗的形式，通过外墙涂料颜色的变化、拼接以及由于功能需求，不同材质的体量穿插，寻求该建筑独具一格的鲜明特色。

建筑内部空间设计注重整体性，克服现状建筑空间低矮凌乱的缺陷，试图营造一个富有时代感、纯净、唯美的办公建筑空间。

Shanghai Expo 2010 Urban Best Practice Area Northern Sub-district UBPA Office Building (hereinafter referred to as UBPA Office Building) is an old building (physic-chemical testing center) renovation project. The existing building was an ordinary multi-storey building with outdated and simple appearance. Except the first floor with a storey height of 4.2m, the storey height was 3.1m for the 2nd - 5th floors and 3.0m for the 6th and 7th floors, which was comparatively low for offices. Additionally, the standard of the public facilities for the building was relatively low, such as lifts and toilets, which have failed to meet the requirement of modern offices. Since the northeast side of the existing building was out of the red line, in order to reduce the occupation of roads, the Employer demanded for making the first floor into an arcade form during renovation. Through renovation, the building will become a permanent office building in the Urban Best Practice Area (UBPA). The building geometry was basically in line with the old building geometry, as a multi-storey building with a "U" plane. The elevations were renovated in line with the concise, economic but nice principle: for the elevation in all directions, windows were arranged equally on the wall; by changing and slicing the exterior coating colors, with different materials as per functional demand, it was to seek a unique character for the building upon renovation. The internal space of the building was design with a focus on integrity to overcome the defects of the existing building such as shallow space and chaos and to create a modern office building space with purity and beauty.

改造前外景 / Exterior view before renovation
改造后外景 / Exterior view after renovation

BETTER CITY BETTER LIFE

总平面图 / Master plan
首层平面 / The 1st floor plan

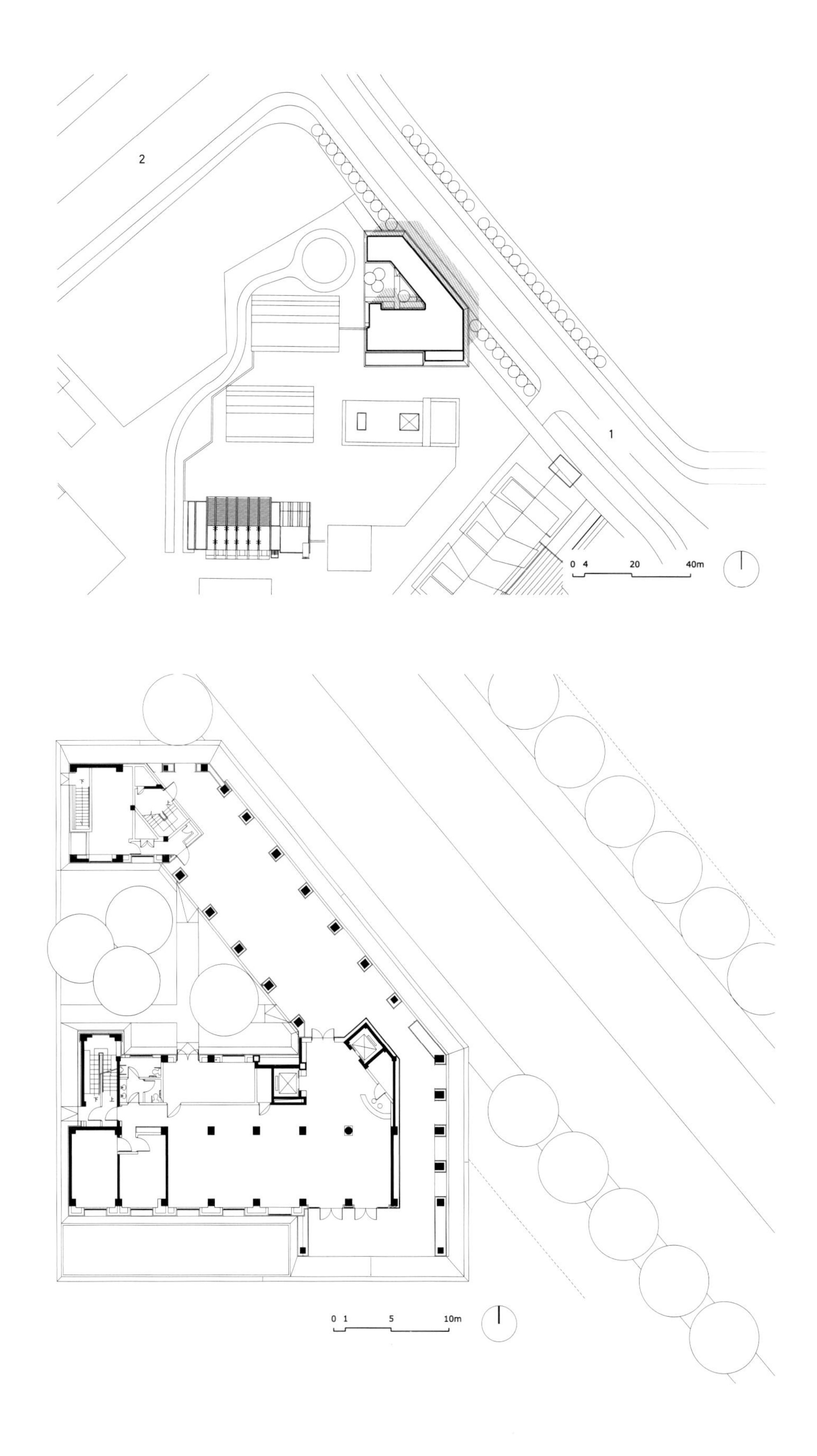

1 南车站路 / South Station Road
2 中山南路 / Zhongshan South Road

标准层平面图 / Plan for the standard floor
外景 / Exterior view

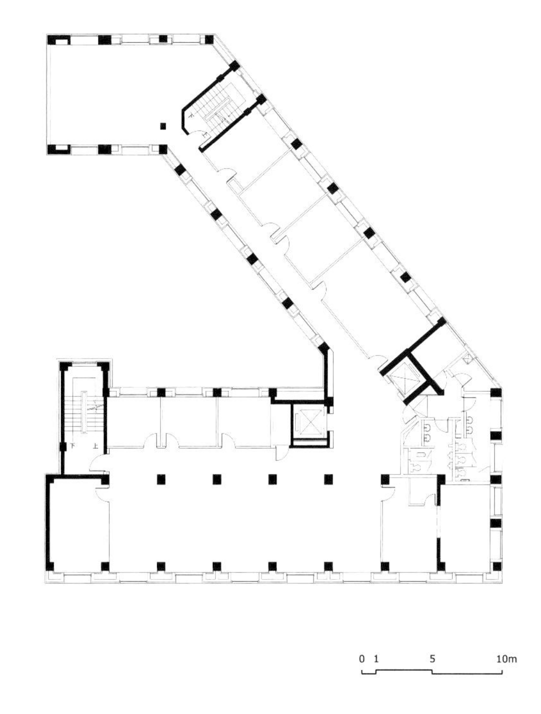

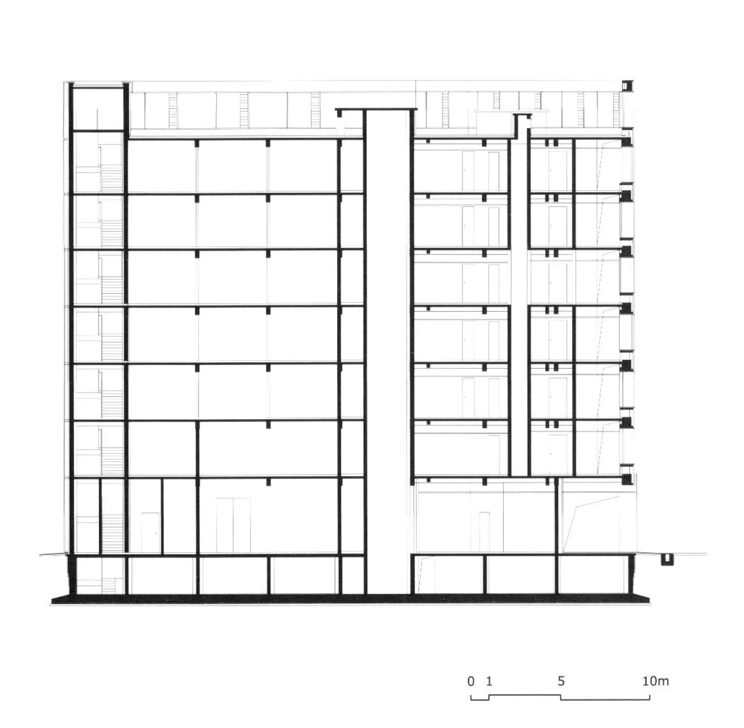

外景 / Exterior view
局部立面 / Partial elevation
剖面 / Section

内景 / Interior view
外景 / Exterior view

外景 / Exterior view

外景 / Exterior view

北京建筑工程学院新校区学生综合服务楼　北京

BUCEA New Campus Student Comprehensive Service Building, Beijing

2011

该建筑位于北京建筑工程学院大兴新校区内，是学生宿舍区内的一个小型公共建筑，它面临的最主要的问题是功能转变。该建筑在校园建设的初期，将作为宿舍区的配套商业设施，校园基本形成后其功能又将转换成一个多功能建筑。就此校方提出了多种设想：多功能厅、展厅、活动中心、室内体育馆……针对这种情况，我们将功能的自由转换作为设计构思的出发点，并制定了下列设计要点：一层无柱；单元式模块组合；电气、设备为功能转换提供有利条件；内部夹层采用较易拆除和可回收的材料；增加节能环保设施。

根据设计要点，我们提出了一个由 10 米 ×10 米单元排列形成的 60 米 ×60 米的正方形平面，每个单元由中间的一个天窗和四坡屋顶组成。

在这个基本形状下，结合节能要求提出几个重要措施，最后形成了现在的造型。

措施 1：由于该建筑是一个单层建筑，因此屋顶的保温隔热是一个不好解决的问题。因此我们在四面坡的屋顶上又增加了一层反向四面坡屋顶，同时将 2 米高的大梁隐藏在两层屋顶之间。

措施 2：中间天窗采用可开启的方式以便通风。

措施 3：在建筑外侧设置了一层外廊，为外侧的大窗户提供了遮阳设施，同时也为店铺和学生活动提供了一个半室外空间。外廊的另一个重要的功能是为大跨度的梁提供了支撑。另外，为了丰富外立面，将外廊处理成在轴网旋转后将外侧单元进行切割的形式。

该建筑为钢筋混凝土结构，为了表达建筑几何的逻辑关系，屋顶、切割面为清水混凝土，切口内为木质外墙。

The building located on Daxing Campus of Beijing University of Civil Engineering and Architecture (BUCEA) is a small-scale public building. The major problem was the change in functions. The building would be used as the auxiliary commercial facilities for the dwelling area in the early period of campus construction and converted into a multi-function building upon substantial completion of the campus. In this regard, different assumptions were proposed: multi-function hall, exhibition hall, activity center, indoor gym, etc.. For this, we took the free conversion of functions as the starting point for the design concept and formulated the following design keynotes: 1st floor: column-free; combination with unit modules; favorable conditions for functional conversion by electrics and equipment; easy-to-remove and recyclable materials for the internal interlayer; addition of energy-saving and environmental-friendly facilities. According to the design keynotes, we proposed a square plane of 60m×60m arranged with 10m×10m units, while each unit is composed of one central skylight and four-slope roof. Based on the above, in combination with the energy-efficient requirements, several important measures were proposed before the current model was eventually formed.

外景 / Exterior view

外景 / Exterior view
西立面图 / West Elevation
剖面 / Section

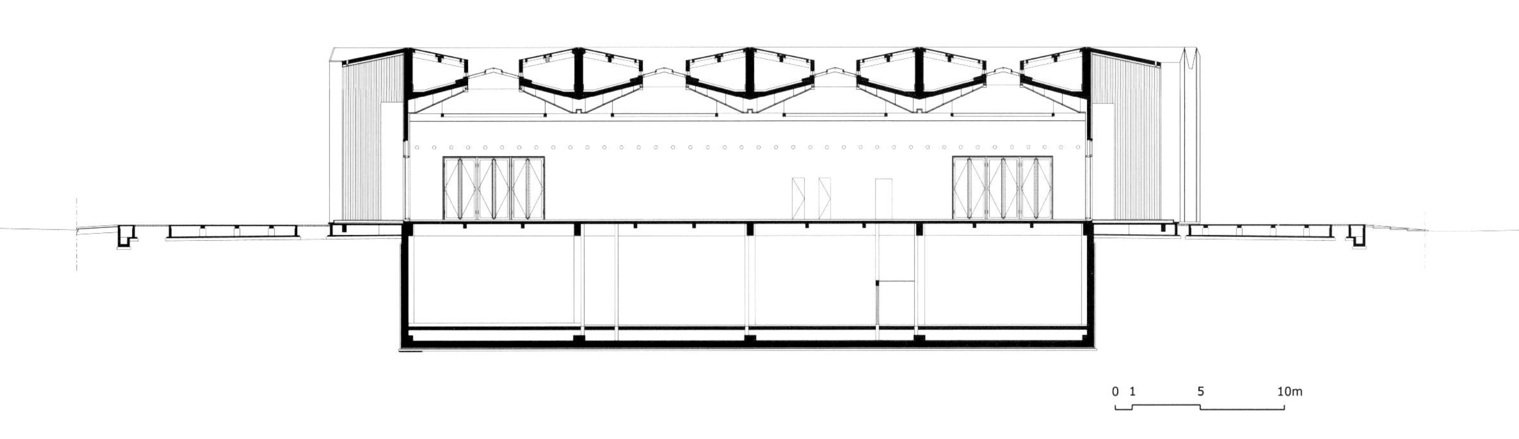
0 1 5 10m

总平面图 / Master plan
外景 / Exterior view

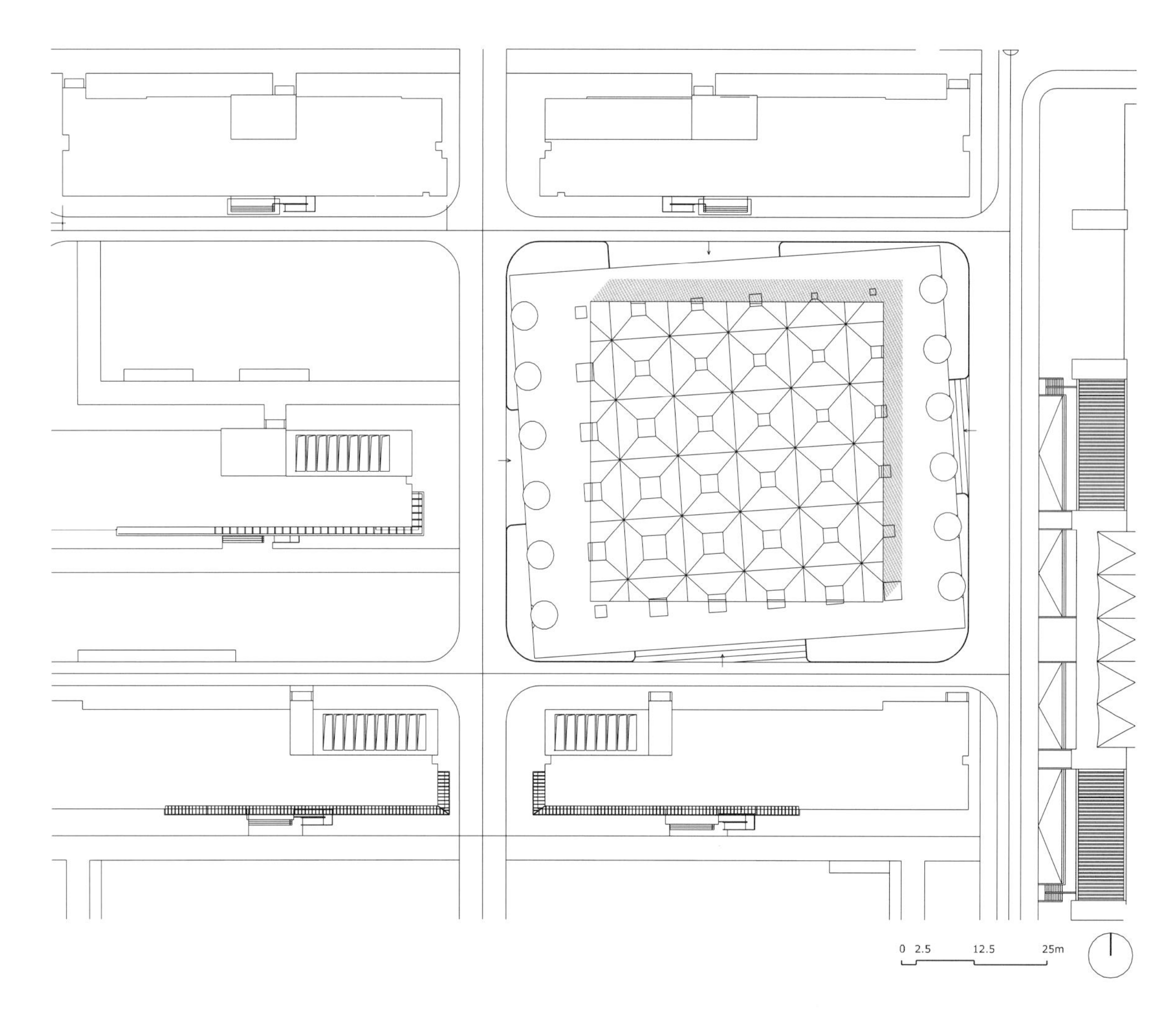

首层平面 / The 1st floor plan
外景 / Exterior view

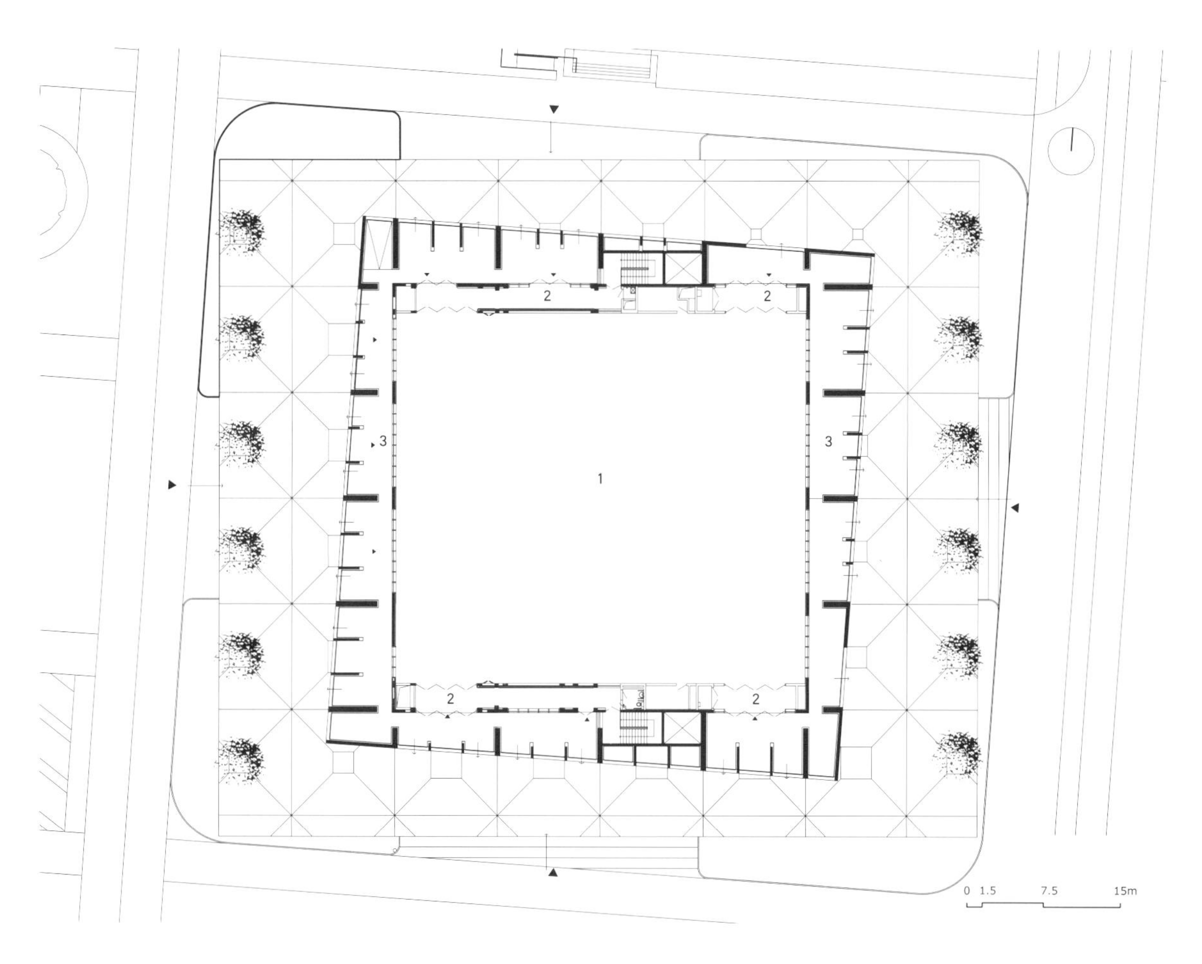

1 多功能厅 / Multi-function hall
2 门厅 / Entrance lobby
3 外廊 / Verandah

外景 / Exterior view
内景 / Interior view

Measure 1: since it was a single-storey building, the insulating layer for the roof was a hard nut. For this, on the four-slope roof, we designed one additional concrete layer of a reversed four-slope roof and hid the beam of 2m-high between the two roofs.

Measure 2: the central skylight was opened for ventilation.

Measure 3: one verandah was arranged outside the building to shelter the exterior windows as well as provide a semi-exterior space for shops and activities of students. The verandah is also functioned as a support for the large-span beam. To enrich the exterior elevation, the verandah was treated to cut off the exterior units upon rotating the axial net.

The building is of a RCC structure. To express the geometric logic relation of the building, concrete was applied on the roof and cut surface, while the wooden exterior wall was used inside the cut.

外景 / Exterior view
分析图 / Diagram

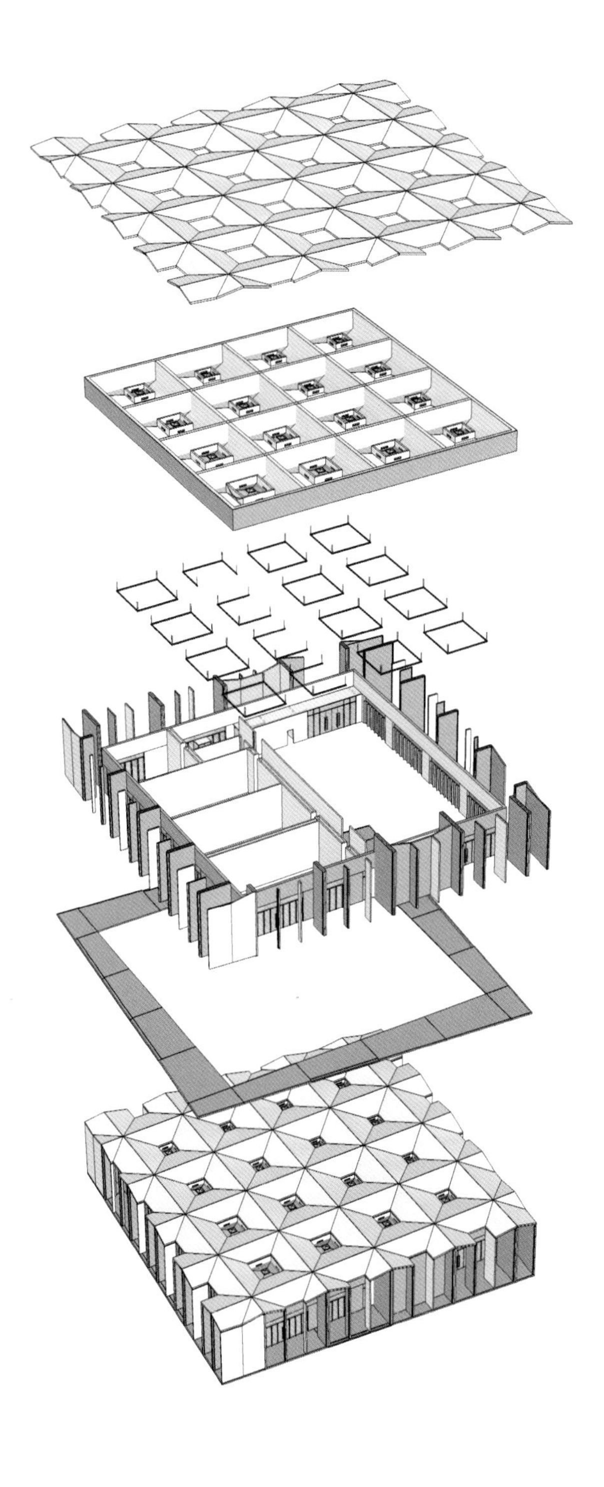

外景 / Exterior view

外景 / Exterior view

内景 / Interior views
外景 / Exterior view

D2 地块由两栋建筑组成，其中体量较大的 A 栋办公楼位置靠近主要道路的交叉路口，可以起到更好的标志作用。

D2 地块位于山顶，自然地形已被破坏，整个场地竖向设计依据用地西南侧环状道路规划标高及西纬六路道路标高进行。从道路标高来看，整个用地高差 8 米，西低东高。在建筑设计时充分利用边界地形条件，通过建筑化手法使自然地形与周围道路、地形很好地衔接，形成一个自然的曲面，就像一张柔软的“毯子”。这张柔软的“毯子”一方面覆盖整个地下空间，包含主入口雨篷，另一方面作为景观平台将两栋建筑主体的纯净体形展现出来。两栋建筑主入口顺应道路坡度与道路衔接，均从地下一层进入，并形成平坦的入口广场，车辆可以直接抵达主入口。

根据鄂尔多斯范式要求，对两栋办公楼的体形进行了长宽高的限制。本方案在满足相应体形模块的范式条件下，做了适当的体形变异，如两个维度的抹角处理，以期与景观平台形成一个完整的整体，在造型上连贯统一并纯净独特。

立面设计结合采光通风设计，A 栋采用了匀质开窗加渐变穿孔铝板及光导管口的处理，寻求一种理性的变化，新颖独特。B 栋采用了匀质深窗洞及变化的穿孔铝板，满足通风遮阳及变化的立面效果。

鄂尔多斯 20+10 项目 D2 地块 鄂尔多斯

Ordos 20+10 Project Lot D2, Ordos

2011

Lot D2 consists of two buildings.Office Building A with a bigger size is close to the intersection of main roads, as a proper symbolic effect. Since Lot D2 is located on top of the hill, with the natural landform already damaged, the entire area was vertically designed according to the elevation for the southwest ring road planning and the road elevation of Xiwei 6 Road. In view of the road elevation, the land has a height difference of 8m, low in the west and high in the east. The architecture design takes full use of the boundary terrain conditions and joins, by architectural method, the natural landform properly with the surrounding roads and site terrain to form a natural curve, just like a soft "carpet". The "carpet" covers the whole underground space, including the main entrance shelter, and is used as an outlook to display the pure geometry of the two main buildings. The main entrance of the two buildings links to the road in compliance with the road slope, for access from the basement, and forms a flat entrance square from which vehicles can directly reach the main entrance. According to the paradigm requirement of Ordos, the geometry of the two office buildings is limited by dimensions. On the premise of meeting the paradigm conditions of the relevant geometric modules, the program changes the geometry moderately, such as cornering treatment of two dimensions to integrate with the outlook for a continual and unified, pure and unique model. The elevation is designed in combination with the lighting and ventilation design. Building A is equally opened with windows and treated with gradual-variation perforated aluminum panels for a rational change and novel uniqueness. Building B is arranged equally with deep window opening and changing perforated aluminum panels for ventilation and shading and a changing elevation effect.

外景 / Exterior view

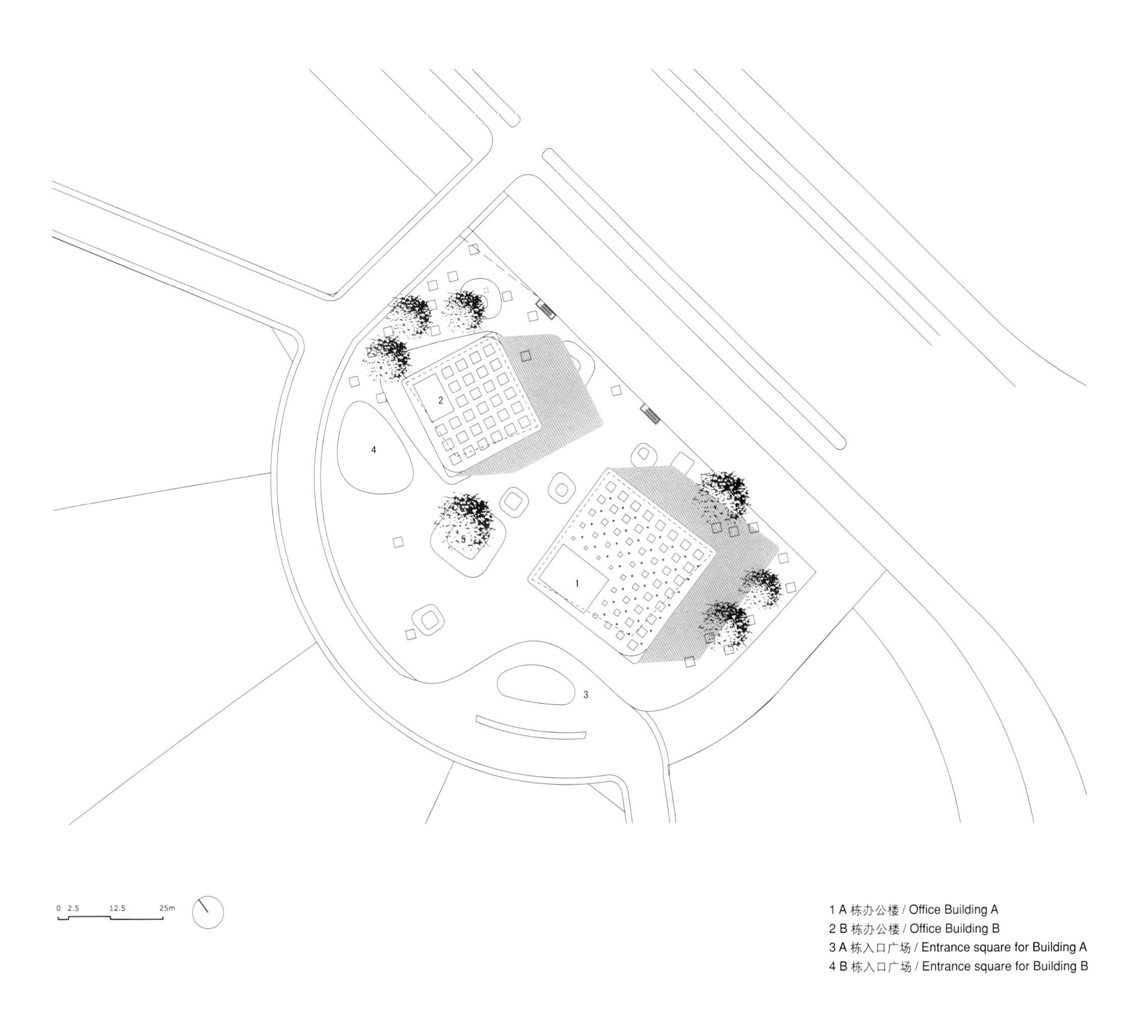

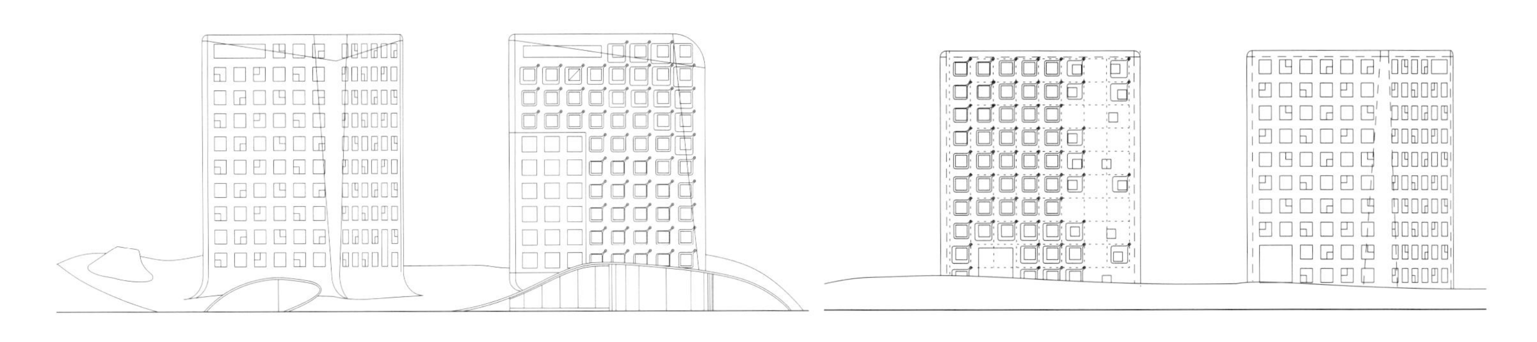

0 2.5 12.5 25m

总平面图 / Master plan
南立面图 / South elevation
北立面图 / North elevation
内景 / Interior view

首层平面 / The 1st floor plan
二三层平面图 / Plan for the 2nd and 3rd floors
四－六层平面图 / Plan for the 4th－6th floors
七八层平面图 / Plan for the 7th and 8th floors

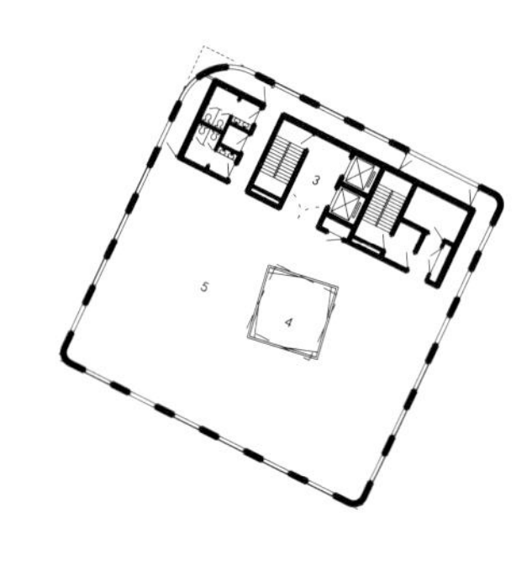

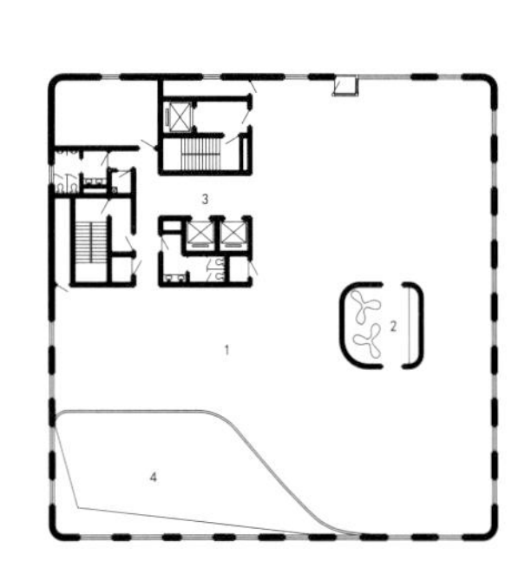

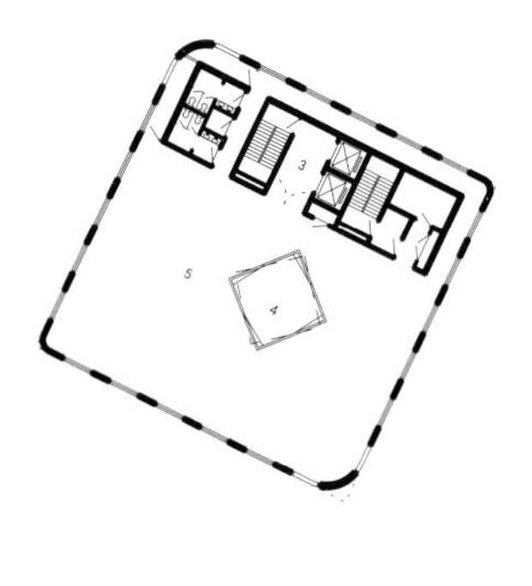

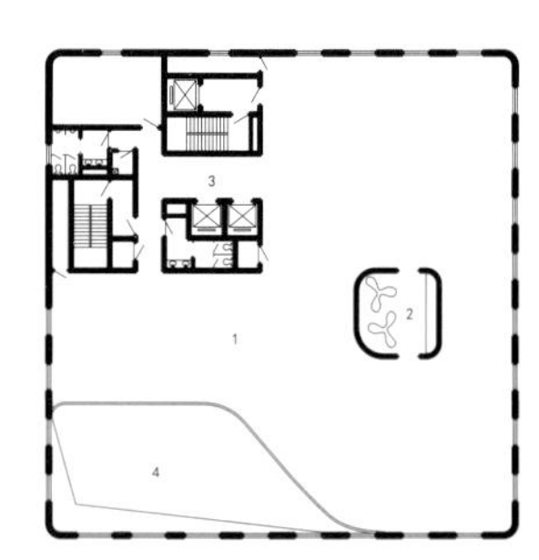

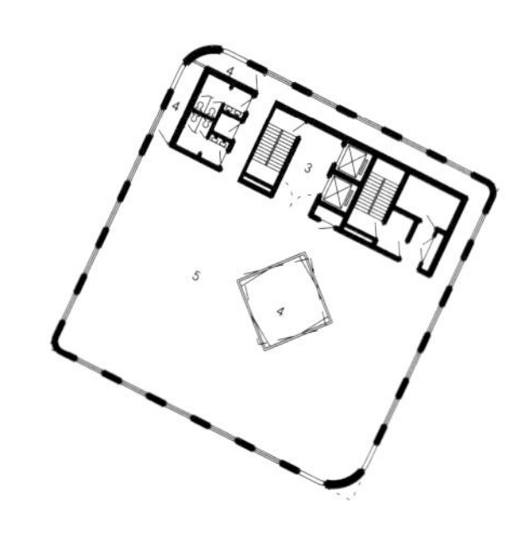

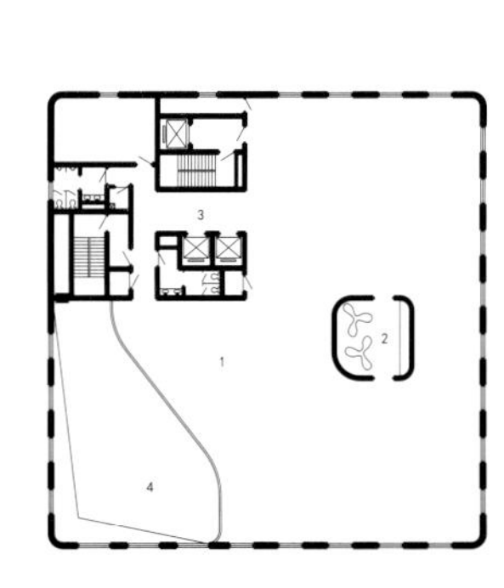

0 3 15 30m

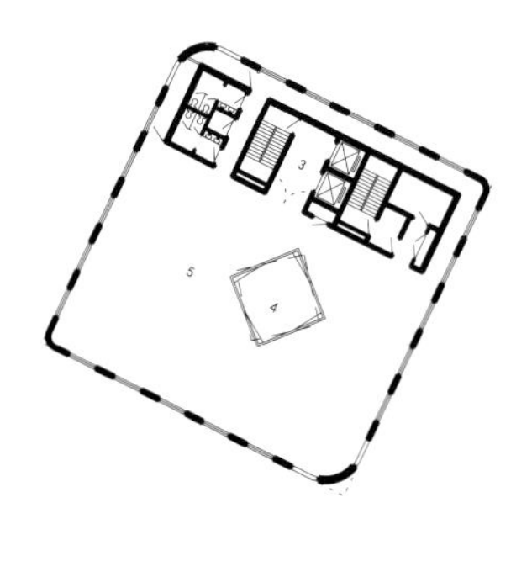

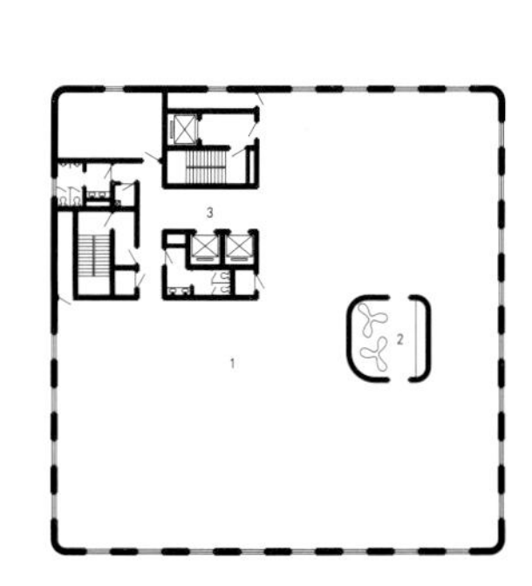

1 A 栋开敞办公空间 / Open office space of Building A
2 休息厅 / Lounge
3 电梯厅 / Elevator hall
4 中庭上空 / Atrium
5 B 栋开敞办公空间 / Open office space of Building B

顶层平面图 / Top floor plan
屋顶平面图 / Roof plan
外景 / Exterior view

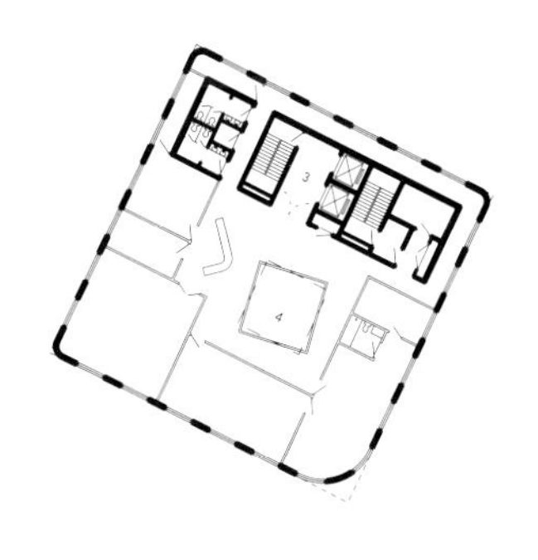

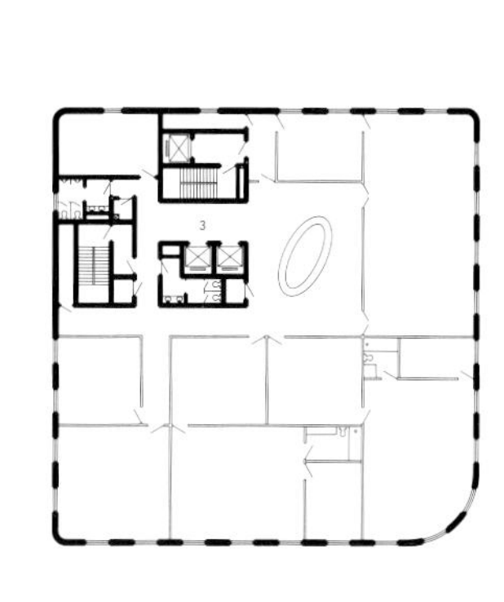

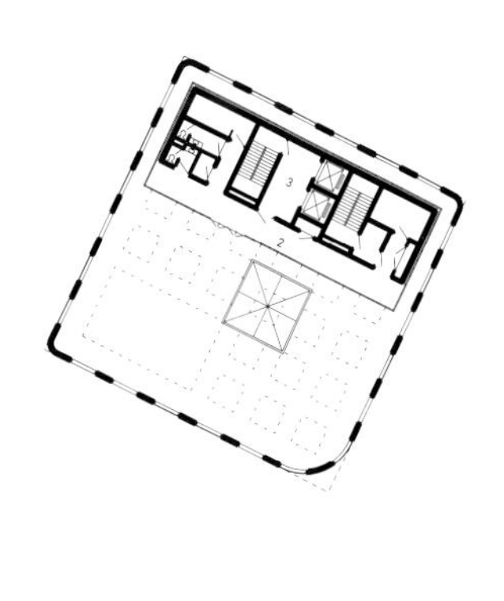

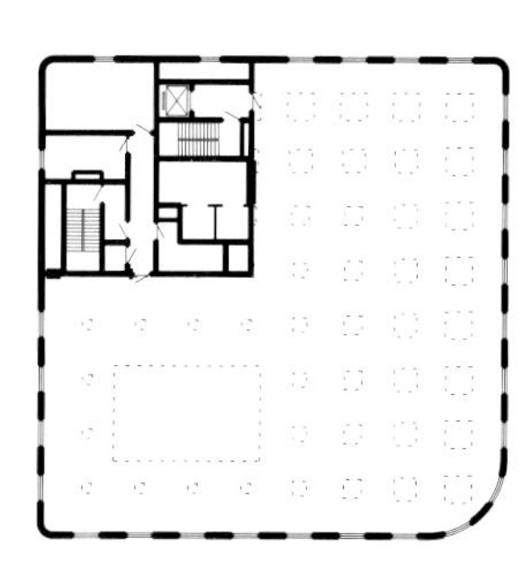

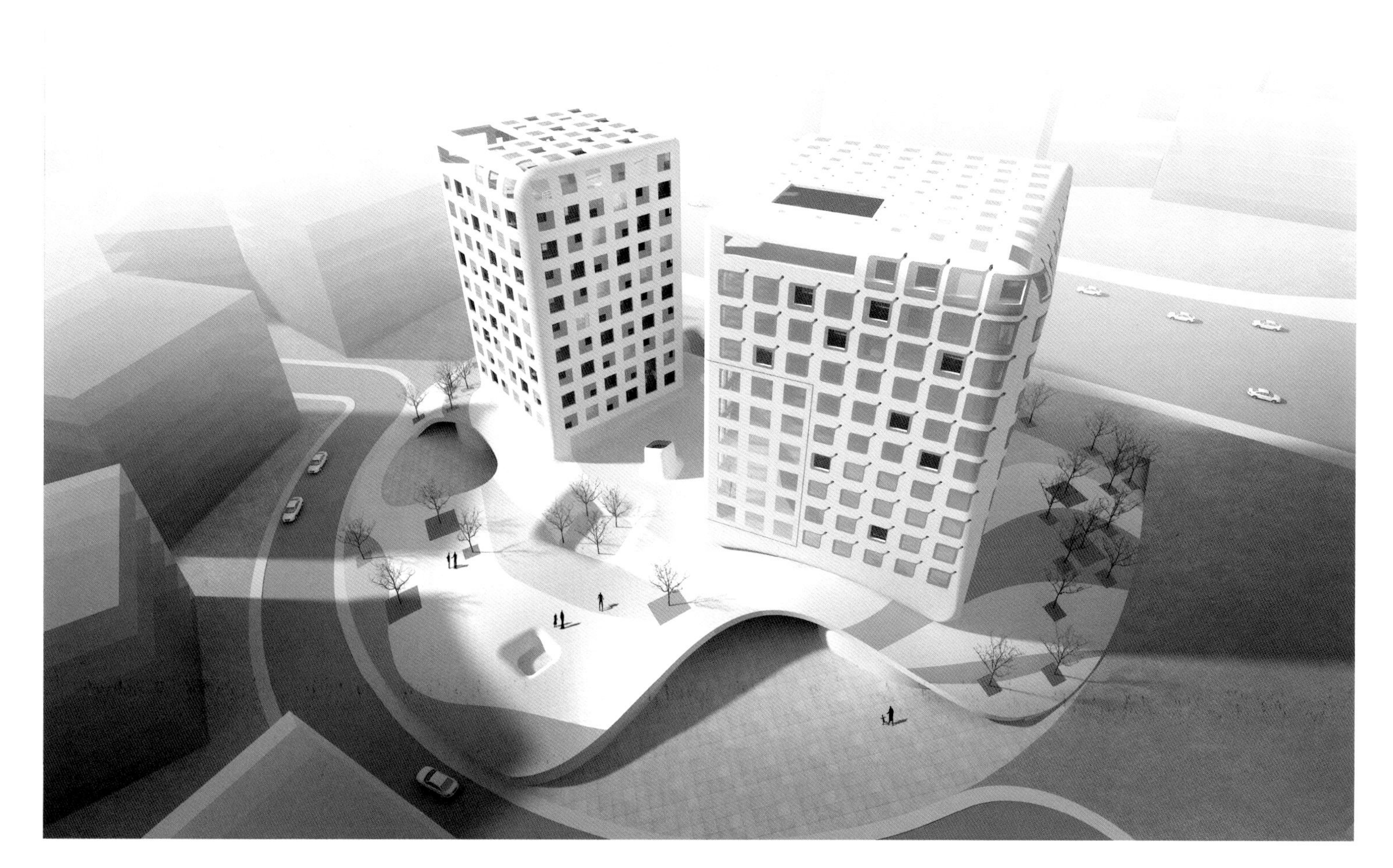

外景 / Exterior view

剖面图 / Section
外景 / Exterior view

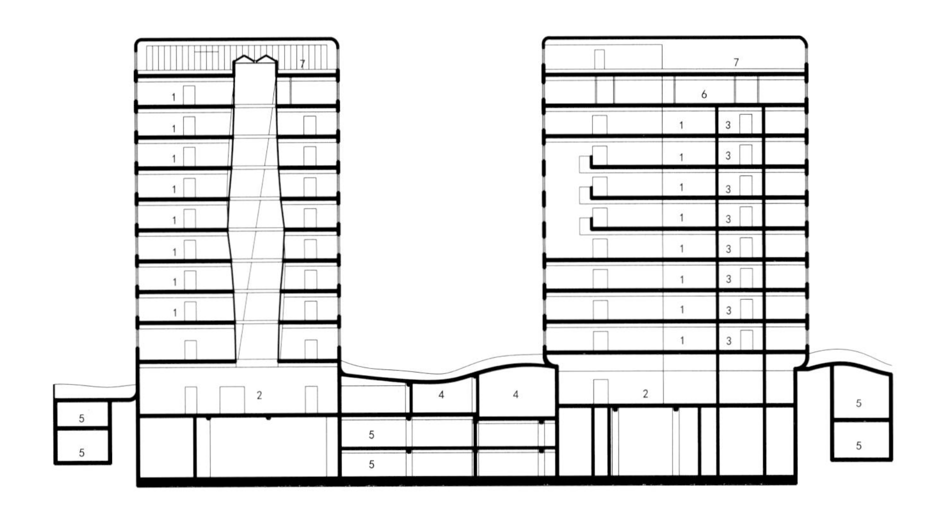

1 开敞办公空间 / Open office space
2 大堂 / Lobby
3 休息厅 / Lounge
4 咖啡厅 / Cafe
5 车库 / Garage
6 董事长办公室 / Chairman office
7 屋顶平台 / Roof platform

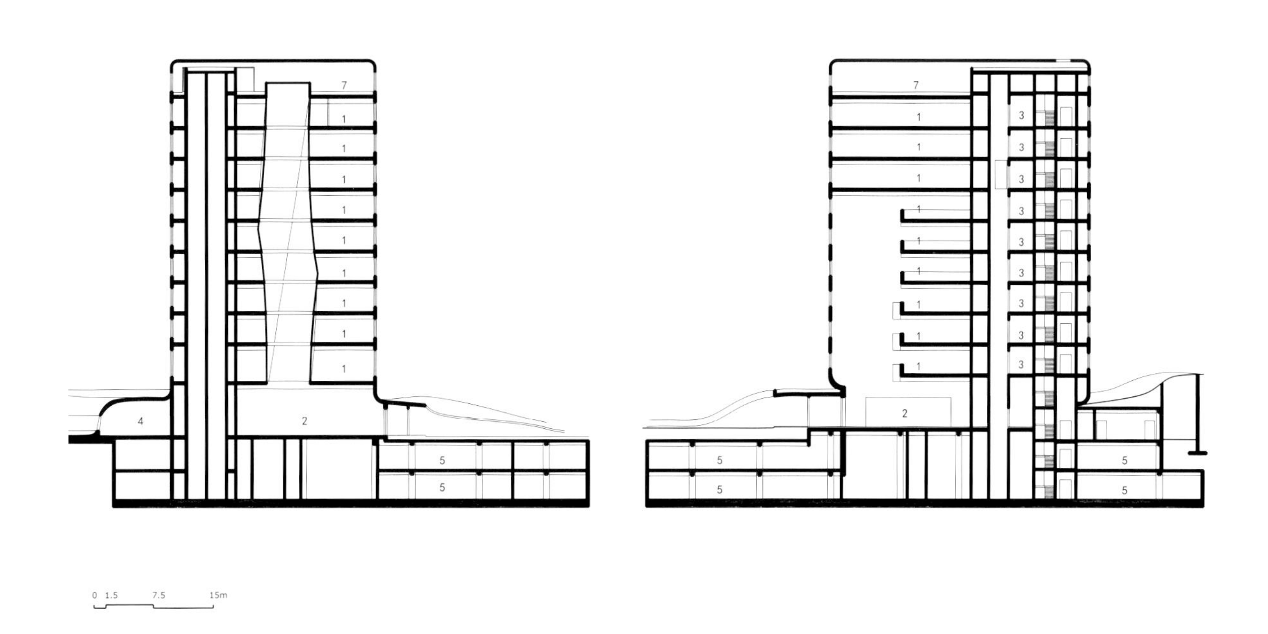
7
1
1
1
1
1
1
1
1
1
4
2
5
5
7
1
1
1
1
1
1
1
1
1
3
3
3
3
3
3
3
3
3
2
5
5
5
5
0 1.5 7.5 15m

外景 / Exterior view

杭州奥体中心体育游泳馆　杭州

Hangzhou Olympic Sports Center Natatorium, Hangzhou

2009

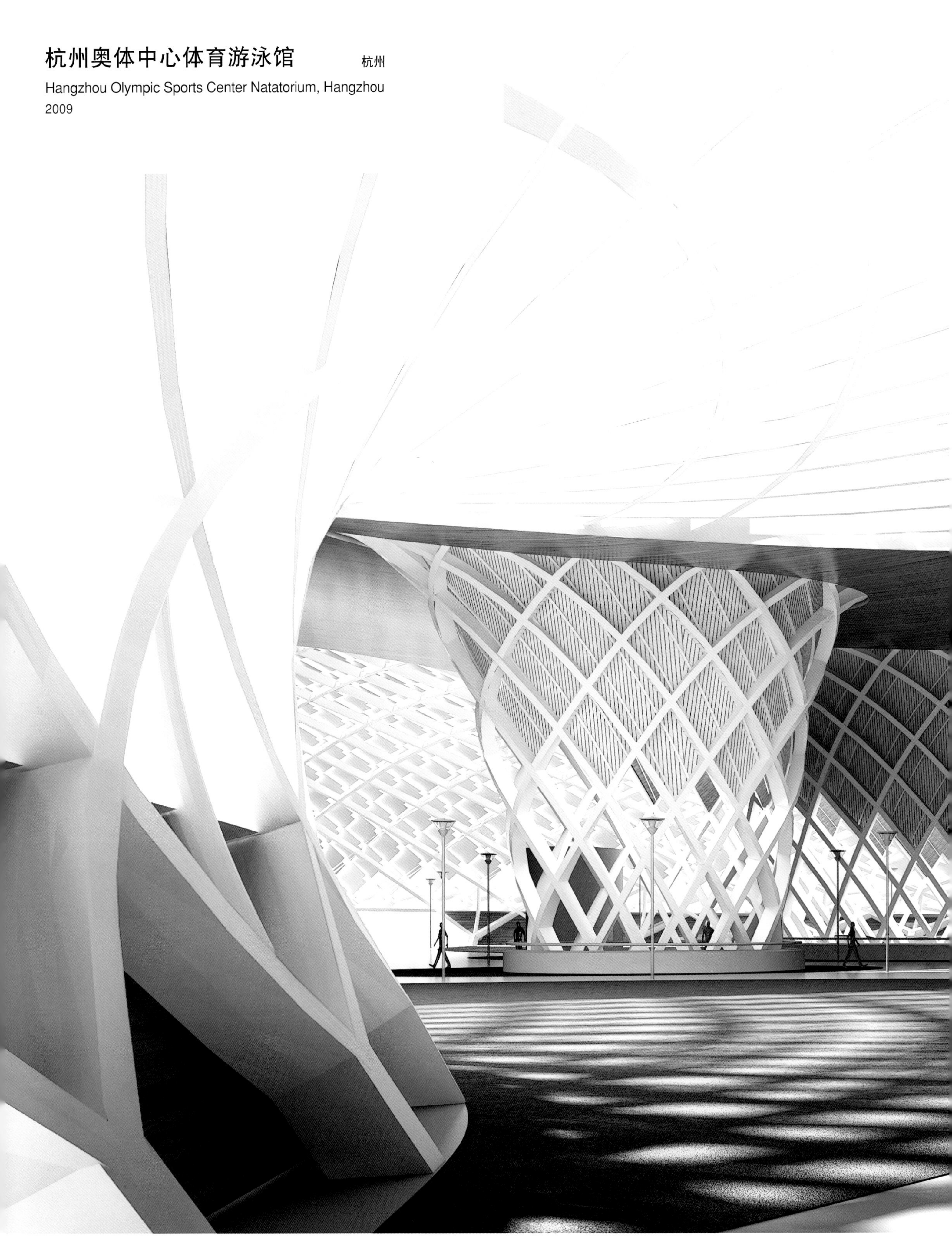

杭州奥体中心体育游泳馆位于奥体中心北侧，占地 22.79 公顷，总建筑面积 396950 平方米。地上主要包括体育馆、游泳馆、商业设施三大部分内容。建筑地下主要包括地下商业设施、机房及地下停车库，可以提供约 2512 个停车位。

体育游泳馆区域内规划是以整个杭州奥体博览中心大区域规划为前提，遵循园区整体规划的结构要求，充分考虑了城市关系、用地特点、控规要求、建筑功能等因素，确定了体育馆与游泳馆两单体建筑沿七甲河顺序布排，形成沿青年路一侧为主要出入疏散方向，沿七甲河一侧为主要商业区域，沿钱塘江一侧为主要景观区域的总体布局。

体育游泳馆周边设置了大量的人员疏散、景观以及商业需求的各类广场设施，在满足赛时大量人员疏散要求的前提下还可以在平时为群众提供一个进行文化活动和休憩的室外空间。

用地内广场纵向划分为三部分，包括建筑四周的地面广场、8 米高的平台广场以及连接地面广场和平台广场的景观草坡。地面广场沿青年路一侧为观众集散区域，并在青年路上设有两个机动车道的开口，供车行使用。沿七甲河一侧的地面广场主要为商业服务，同时为七甲河沿岸的休闲娱乐提供必要的空间。草坡主要集中在沿青年路和七甲河方向，部分草坡朝向钱塘江。沿青年路方向的草坡中植入疏散楼梯，服务于体育馆和游泳馆内观众的室外疏散，其余部分草坡主要起到景观和休闲作用，局部亦可作为应急疏散。8 米平台广场沿建筑周边设置，为体育馆和游泳馆观众服务，并沿建筑设置了一周的景观水池，以保持观众人群和建筑外墙之间的适度距离。另外在建筑的西南侧布置了部分下沉广场，形成地下商业的室外空间，提供必要的采光通风以及消防疏散需求。

Hangzhou Olympic Sports Center Natatorium is located on the north of the Olympic Center, with a land area of 22.79 hectare and a total floorage of 396,950m^2. The superstructures include three major parts: gymnasium, natatorium and commercial facilities. The basement includes mainly commercial facilities, machinery room and garage with 2,512 parking spaces.

The area of gymnasium and natatorium is planned according to the general planning for the entire Hangzhou Olympic and Expo Center. In line with the general planning requirement of the park, adequate consideration is given to urban relation, land use features, control planning requirements, architectural functions and other factors. The two units of gymnasium and natatorium are arranged sequentially along Qijia River to form a general layout with along the main access Qingnian Road, the main commercial area along Qijia River and main scenic area along Qiantang River.

Around the gymnasium and natatorium are a large number of different square facilities for evacuation, landscape and commercial need, which can be used as an outdoor space of recreation and rest for the public while meeting the premise for evacuation.

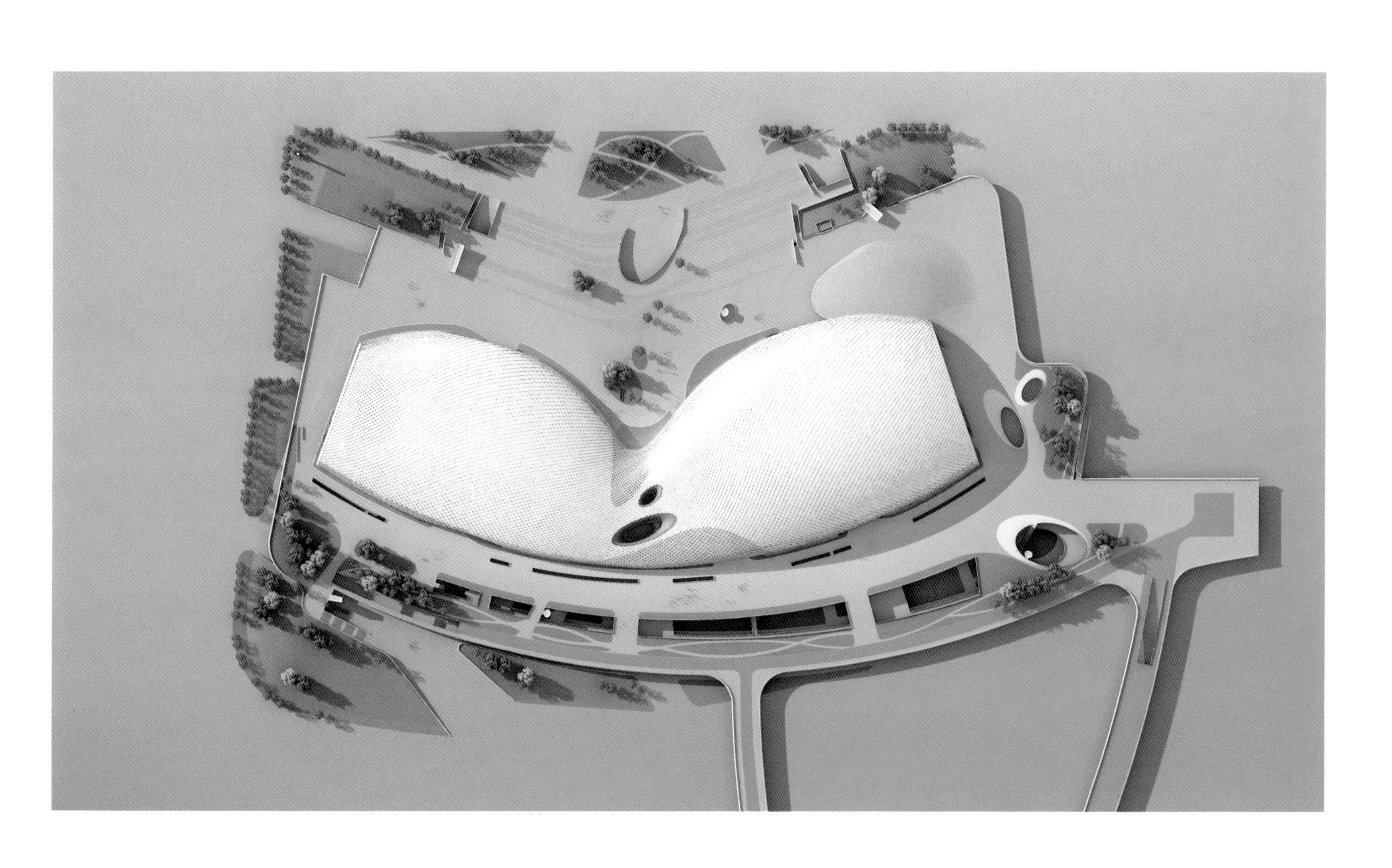

外景 / Exterior view

总平面图 / Master plan
外景 / Exterior view

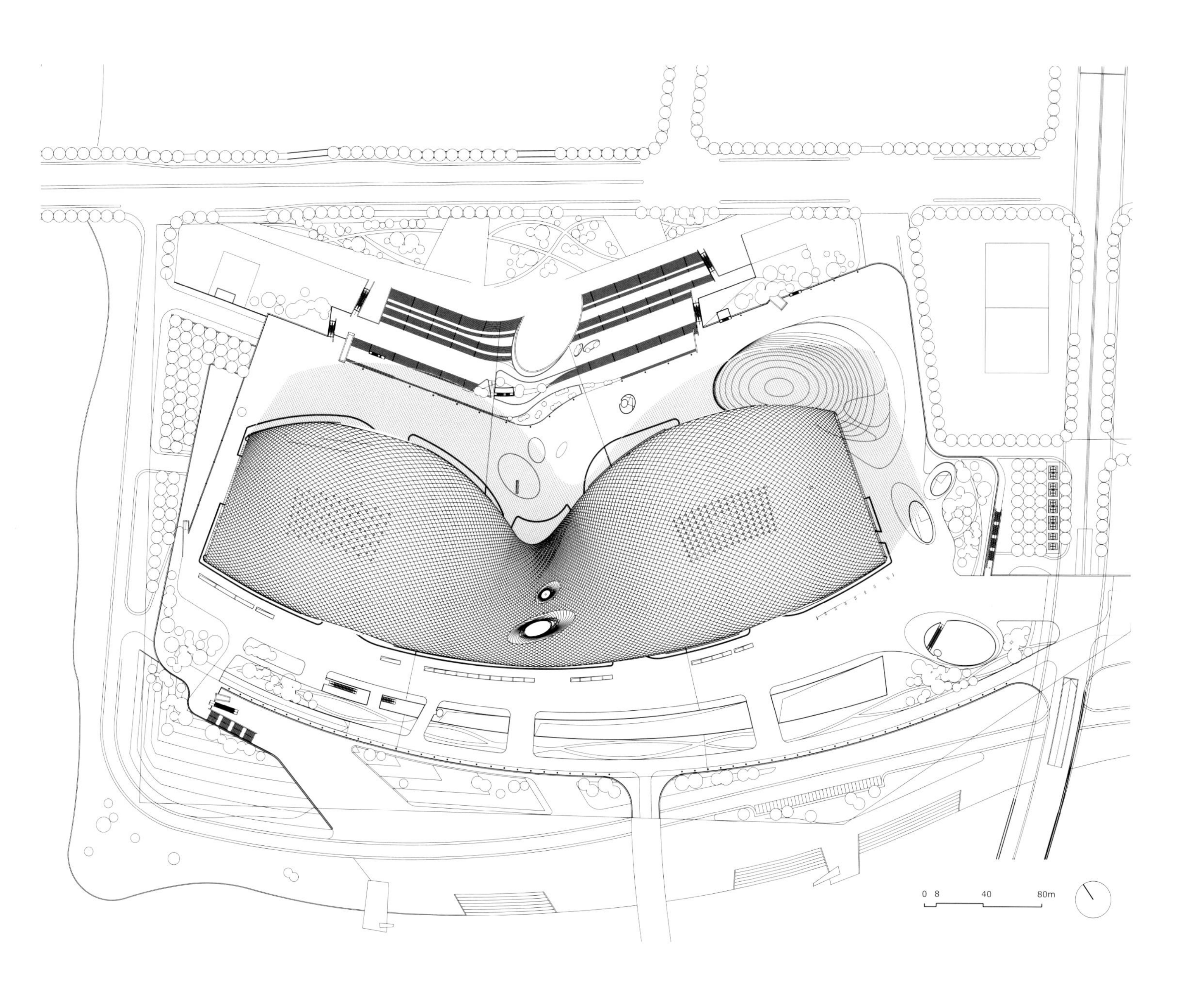

Within the land, the squares are longitudinally divided into three parts, including the ground square around the buildings, platform square of 8m high and the landscape lawn for connection between the ground square and platform square. The side of the ground square along Qingnian Road is a distribution area for spectators, with the two-lane entrance on Qingnian Road for vehicle traffic. The ground square along Qijia River is mainly meant for commercials as well as used as a necessary space for leisure and recreation along the bank of Qijia River. The lawn is mainly concentrated along Qingnian Road and Qijia River, while part of the lane slope faces Qiantang River; the lawn along Qingnian Road is arranged with stairs for outdoor evacuation in service for the spectators of the gymnasium and natatorium. The rest part of the lawn mainly functions as landscape and for recreation, partially as emergency evacuation. The 8m platform square is arranged around the buildings to serve the spectators of the gymnasium and natatorium; along the building, one round landscape water pool to keep a moderate distance between the spectators and the buildings. In addition, on the south of the building, the partially sunken square is arranged to form an outdoor space for the basement commercial facilities to provide necessary lighting, ventilation and fire exit.

二层平面 / The 2nd floor plan
外景 / Exterior view

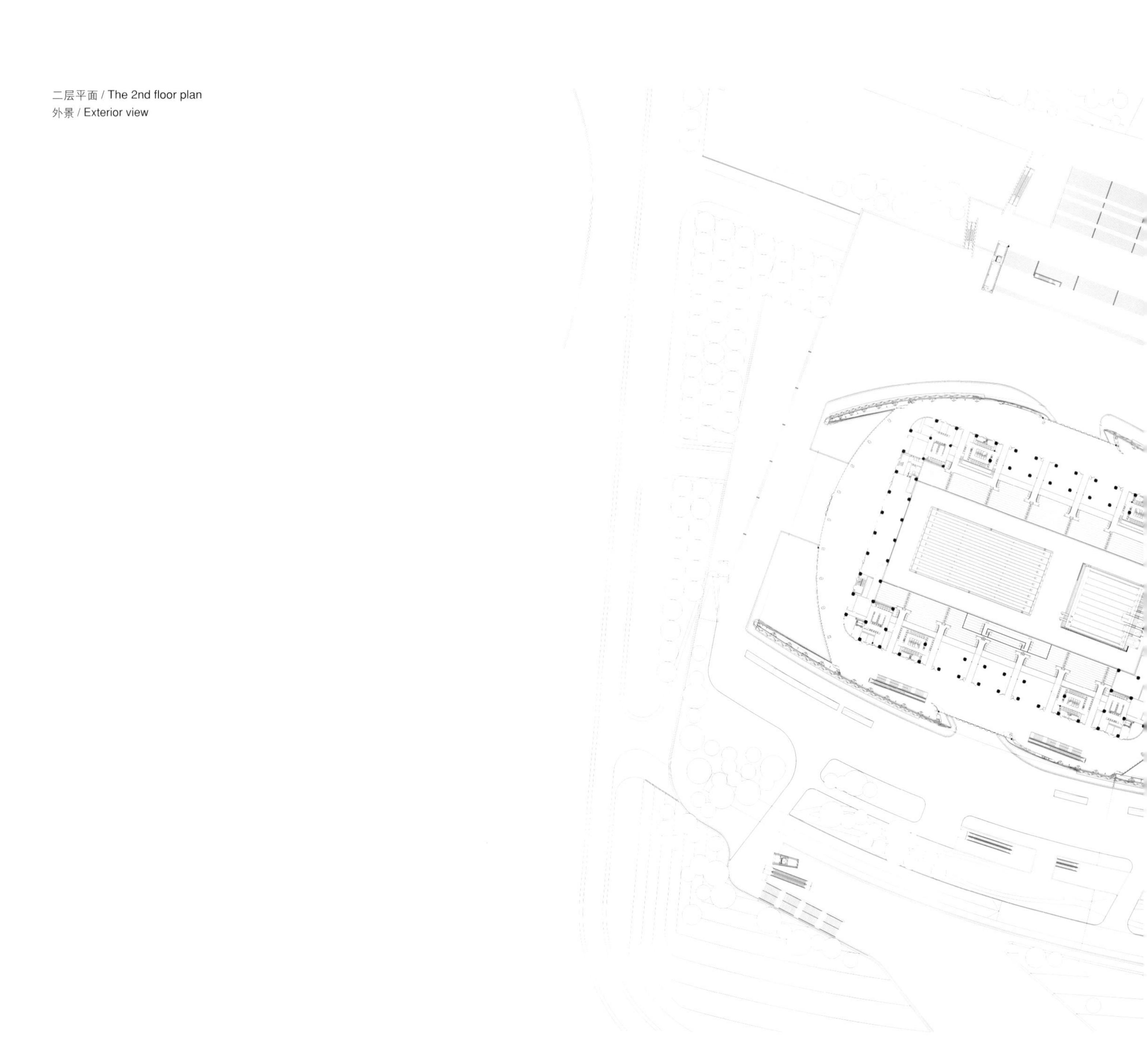

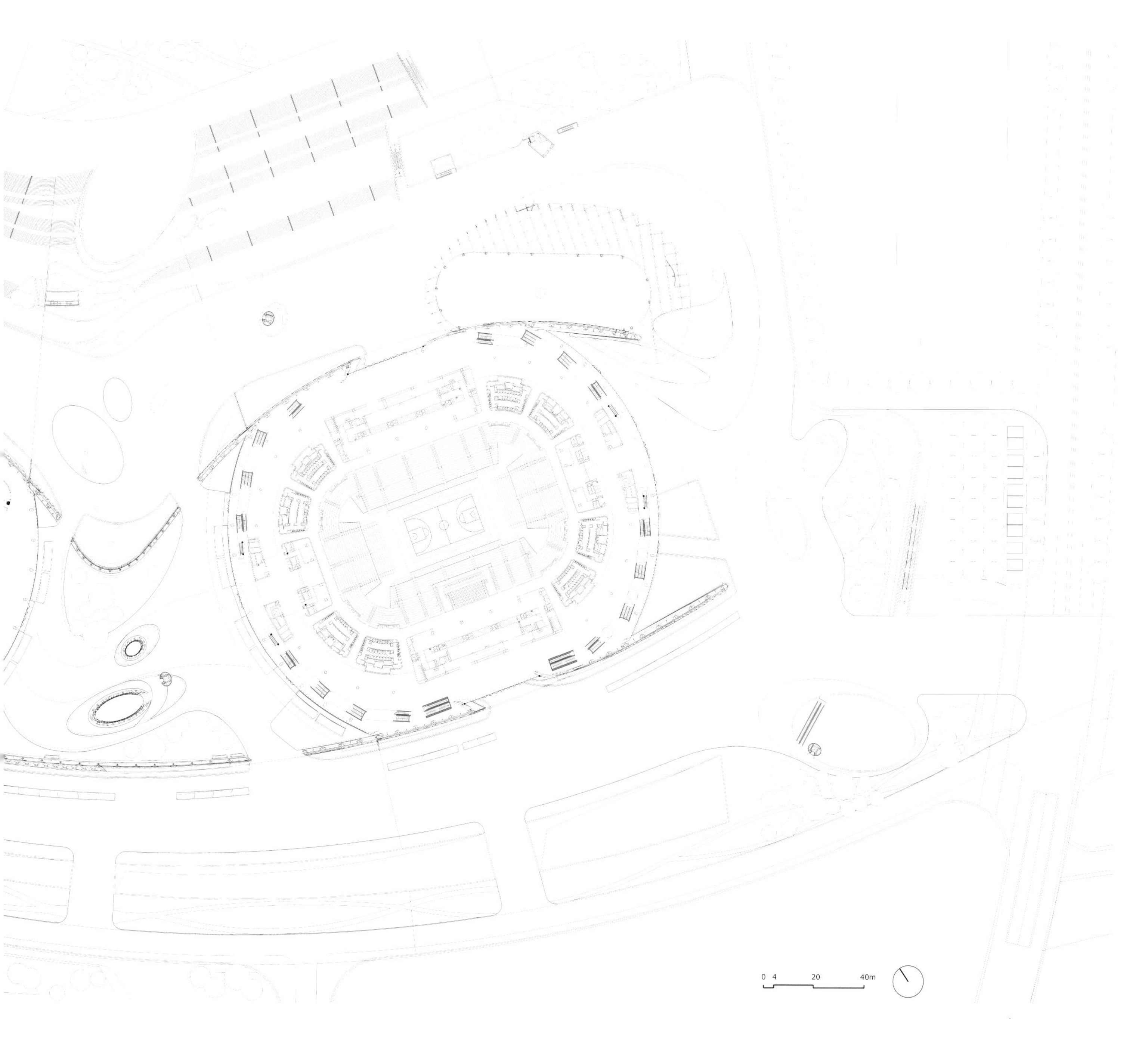
0 4 20 40m

剖面图 / Section
内景 / Interior view

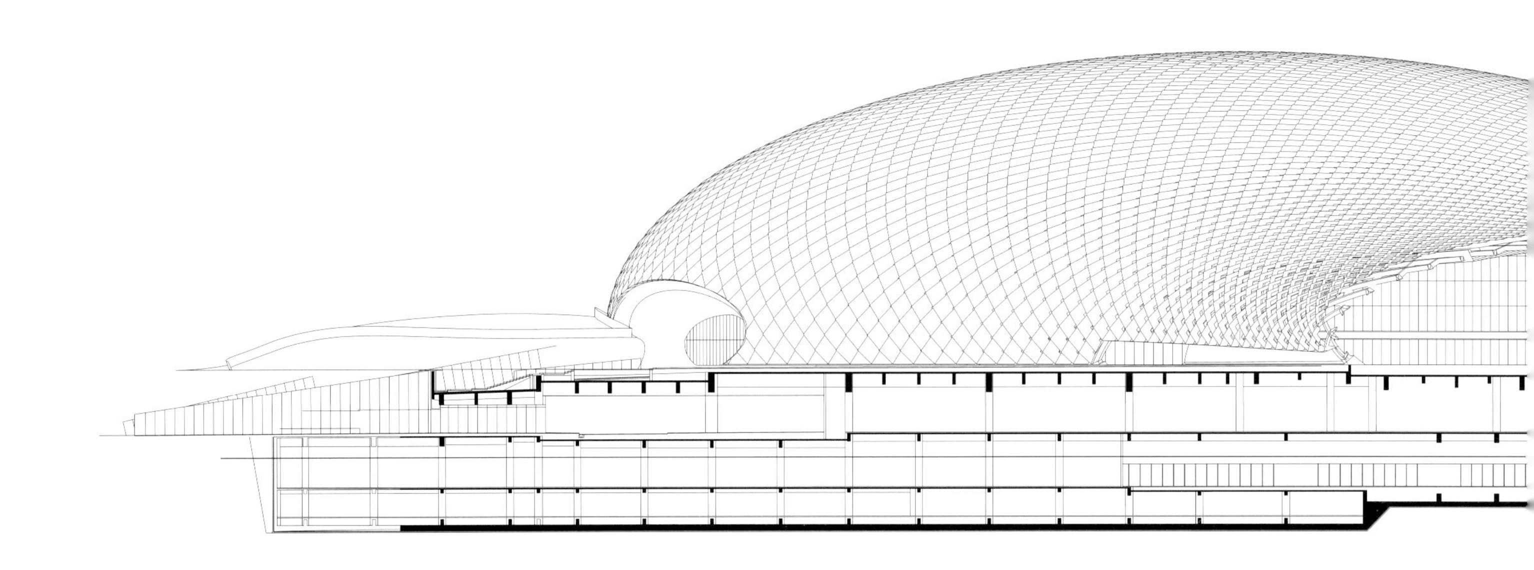

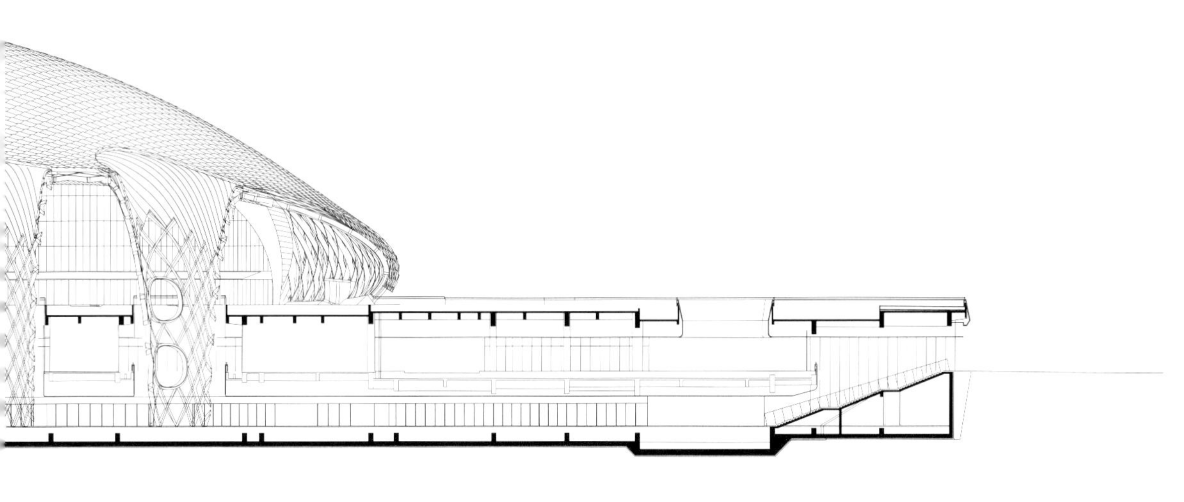

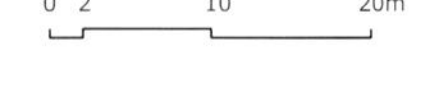
0 2 10 20m

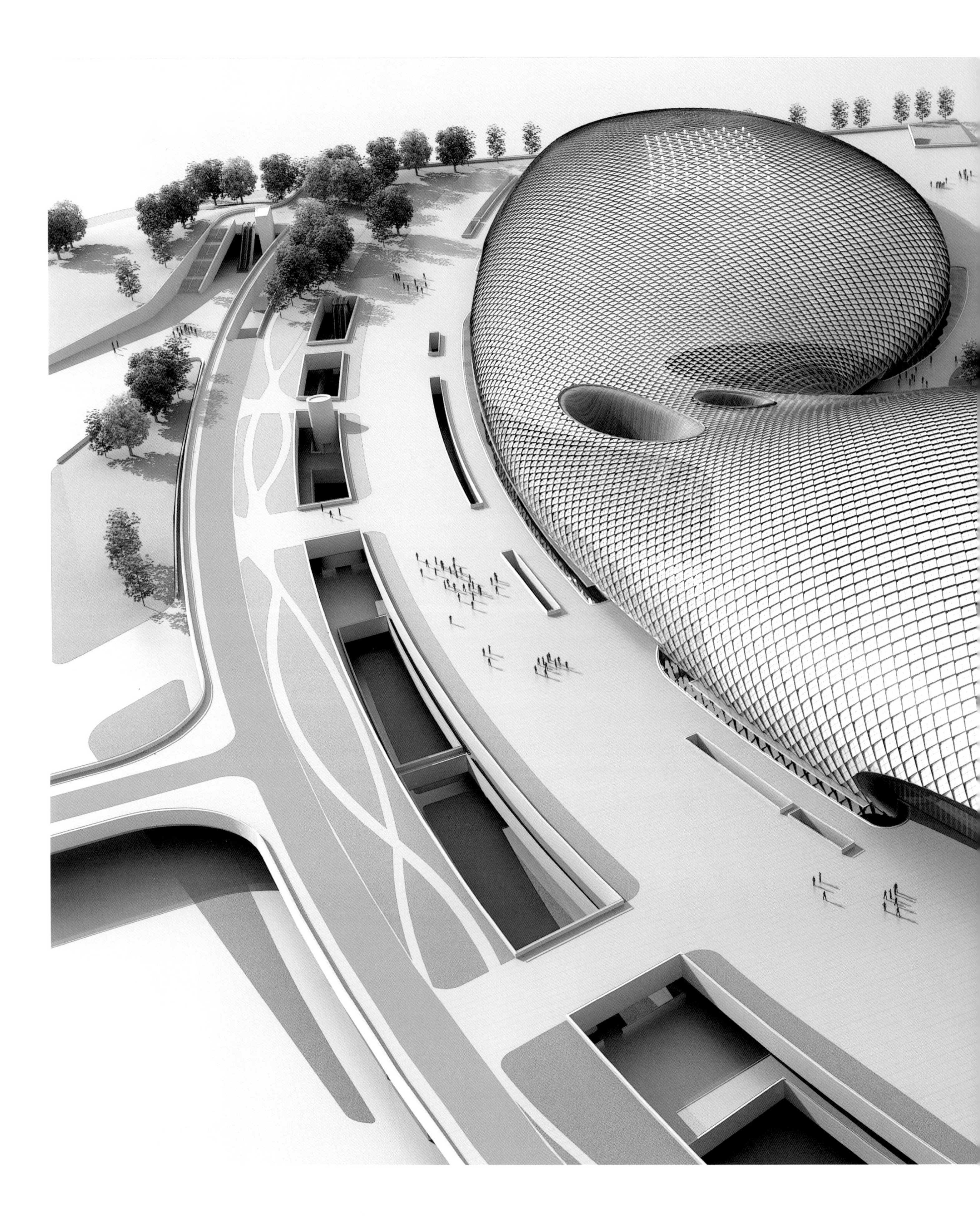

外景 / Exterior view

内景 / Interior view
外景 / Exterior view

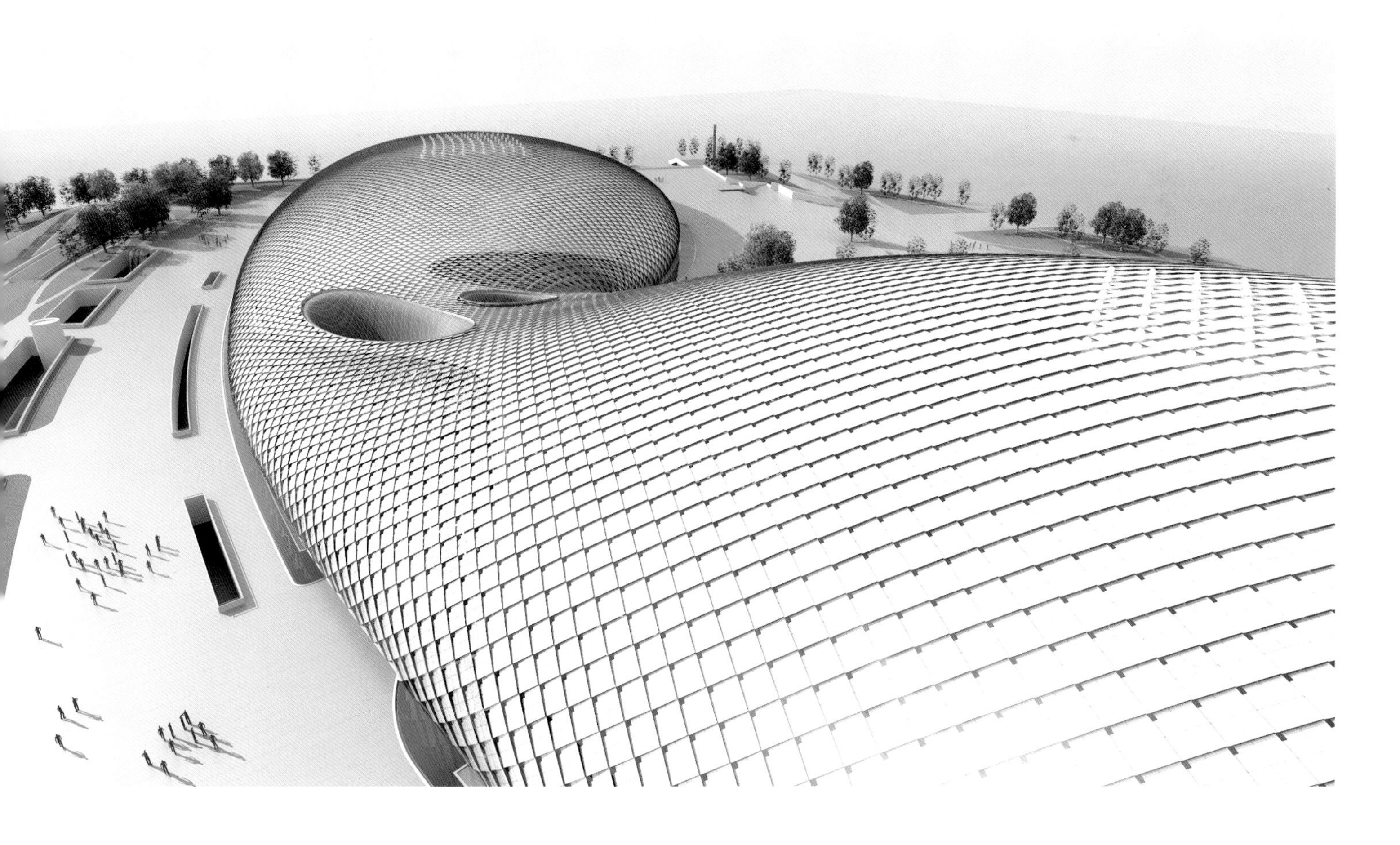

内景 / Interior view

34

外景 / Exterior view

北京延庆设计创意产业园是北京市建筑设计研究院倾力打造的建筑创意产业的综合性交流基地，它位于著名的妫河生态走廊下游，东临延庆县城。园区总体定位为：打造国际一流的集创作、培训、科研、成果展示、文化交流等全面功能为一体的创新型文化创意产业聚落，并形成北京市建筑创意产业及相关行业交流的主流平台。贵宾接待中心位于园区的西南侧，与妫河仅有一片绿地间隔，具有良好的观景视线。

方案的构思首先从功能出发，营造安静、私密的休息环境。首层平面设计成较为封闭的环形空间，一系列的生活功能围绕着中央庭院组织，室内空间均匀连贯，仅依靠家具和木质的隔墙分隔。室内朝向内庭院完全开敞，内庭院安静私密，遍植绿树，满足了人们生理以及心理上对休息空间的需求。

在满足功能的同时考虑与自然环境的融合。首层通过旋转楼梯上到二层平台，完成空间由私密到开放的转换，在朝向较好的方向分别设置了 3 个窗洞，可以看到妫河生态区优美的自然环境。

设计在形式上力求新颖。首层使用厚重的毛石作为室内外的立面材质，二层平台则通过 10 厘米 ×10 厘米的木条编织而成，给主人提供了半私密的室外空间，并带来新颖的空间感受。石材和木材在质感上形成对比颇具乡土建筑的味道，也符合创意产业建筑的性格。

延庆别墅 北京

Yanqing Villas, Beijing

2010

外景 / Exterior view

Beijing Yanqing Design Creativity Industry Park is a comprehensive exchange base of architectural creativity industry made by Beijing Institute of Architectural Design, which is located downstream of the famous Guihe River and adjacent to the county town of Yanqing in the east. The park is generally oriented to make an international first-class village of innovative cultural creativity industry integrated with creation, training, scientific research, achievement exhibition and cultural exchange and establish a mainstream platform of exchange for Beijing architectural creativity and relevant industries. VIP Reception Center is located in the southwest of the park and close to Guihe River at an interval of one green belt only, showing a good view of landscape.

The concept of the project starts with the functions to create a quiet and private leisure environment. The plan of the first floor is designed with a relatively enclosed ring space; livelihood actives are organized in line with the central courtyard; indoor space is well-distributed and constant and is partitioned with furniture and wooden partition walls only. The indoor direction is fully open towards the inner courtyard that is quiet and private and planted with green trees all over to meet the physiological and psychological demands for a rest space.

Consideration is given for integration with the natural environment while meeting the functional requirements. The first floor reaches the second floor platform via the rotating staircase for conversion from private to open space; in the direction of better facing three window openings are arranged from which the beautiful natural environment of the Guihe River Ecological Zone can be viewed.

The design strives to be novel in form. For the first floor, the heavy gross stone is used as the indoor and outdoor elevation materials; the second floor platform is woven with the 10cm × 10cm wooden strips, to give the owner a semi-privacy outdoor space and bring about a novel feeling space; the stone and wood materials are in contrast of texture, with a character of rural architecture as well as creative architecture.

总平面图 / Master plan
内景 / Interior view
外景 / Exterior view

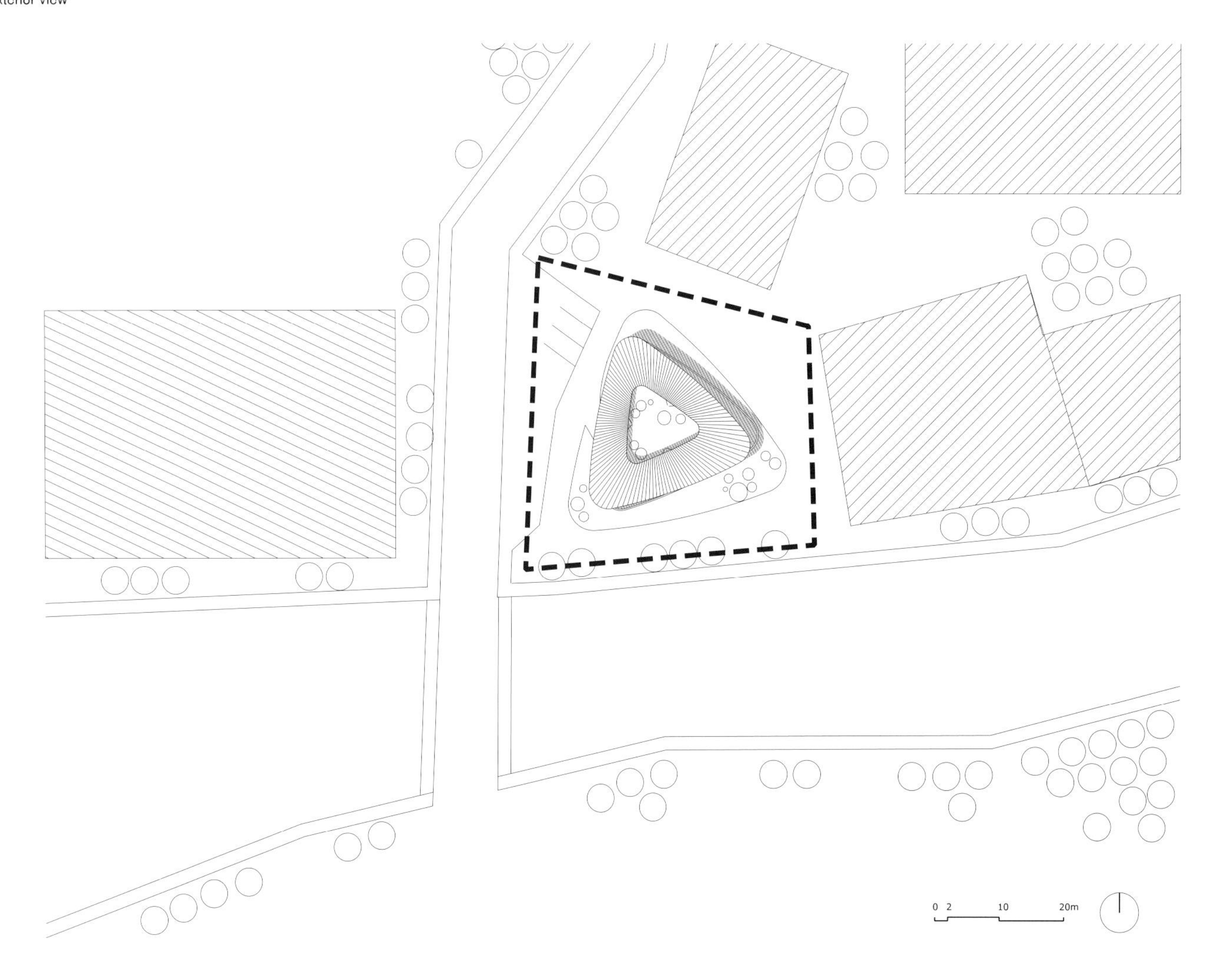

首层平面 / The 1st floor plan
二层平面 / The 2nd floor plan

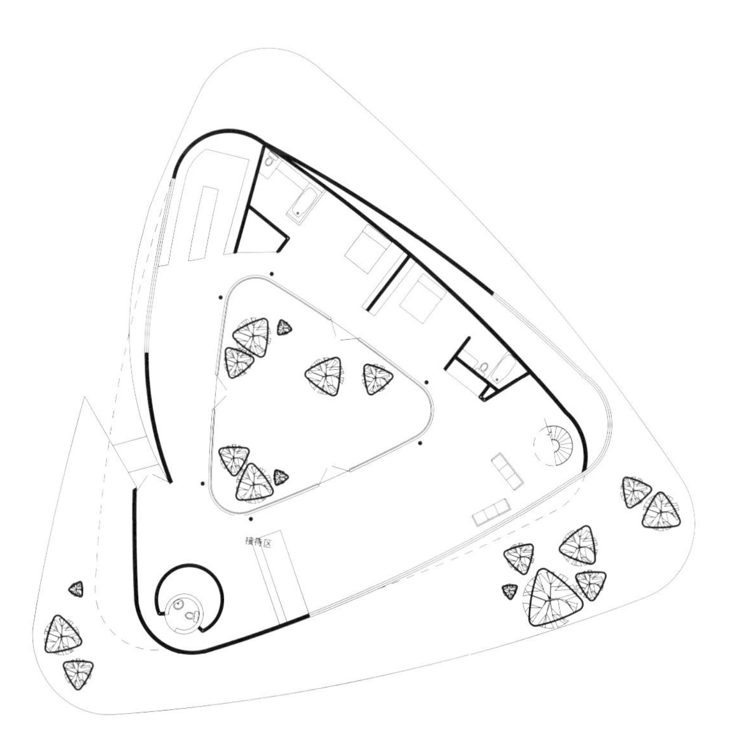

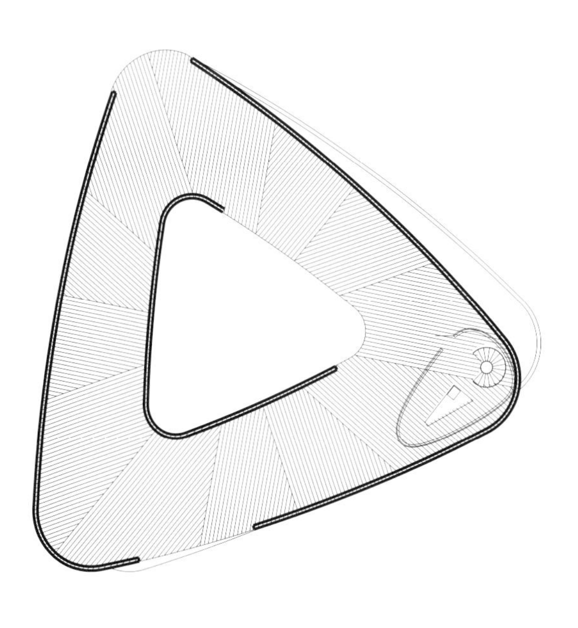

外景 / Exterior view

在北京平谷的一处普通的住宅区内要设计一些联排住宅和少量的双拼别墅。这些别墅的平面是在甲方认可的社会上流行的样式的基础上整合而来的，在整合平面确定后我们再做立面。当时我们向业主表达了一个强烈的意愿，我们希望能把一幢双拼别墅作为一个特案处理，做一点有意思的设计，业主同意了我们的想法。

我们首先对购房者进行了一个预设，即他们应该不会长期居住在这里，来这只是为了换一个环境，当然投资和保值也是购房的另一个主要原因。这样的话，我们就不必按照普通居家的策略设计那种“几室几厅”的住宅。另外，在其周围的那些联排和双拼住宅采用的设计手法均是在用地的中央盖三层高的房子，在用地的前后和一个侧面形成一个外向的U形院子。但中国人传统住宅的模型是形成一个内院，内院会使人有种安逸、舒适的感觉，同时也比较安全。这样我们就在与周围双拼住宅相同的占地内，用一层房子围成一个内院，并根据用地的形状将两户的体量进行了错位处理，于是就形成了这个特殊双拼住宅的基本架构，在房间的布局上，我们兼顾了流动空间和居室的分隔，外墙采用本地的浅色石板条和木格栅。

平谷 3 号住宅　北京

Pinggu No. 3 Residence, Beijing
2011

外景 / Exterior view

It is an ordinary residential community in Pinggu, Beijing, which needs to be designed with some townhouses and a limited number of semi-detached houses. The plans of such villas were designed on the basis of integrating the popular styles acceptable by the Employer before we designed the elevation. At that time, we expressed to the Employer our strong intention to take one semi-detached house for special treat and the Employer agreed on our idea for a meaning design. First of all, we made an assumption of house buyers: they would not reside here for a long time, but only of a change in environment. However, investment and value-preservation is another important reason for them to buy houses. If so, it is unnecessary for us to design a house with bedrooms and living room only according to the common dwelling strategy. In addition, those townhouses and semi-detached houses around were all designed with one three-storey house in the center of the lot. An outward U-type courtyard was created at the front and back and one side of the lot. However, according to the Chinese traditional house model, it is to form an internal courtyard which can make dwellers to feel quiet and comfortable as well as safe. Thus, we fenced an internal courtyard on the lot similar to the surrounding semi-detached houses, with one storey of house. In addition, according to the shape of the lot, the sizes of two units were staggered to form the basic framework for this special semi-detached house. In terms of room arrange, we took into account the partition of flow space and rooms, while the exterior wall was finished with the local light-color stone strips and wooden grids.

总平面图 / Master plan
剖面图 / Section
东立面图 / East elevation
西立面图 / West elevation

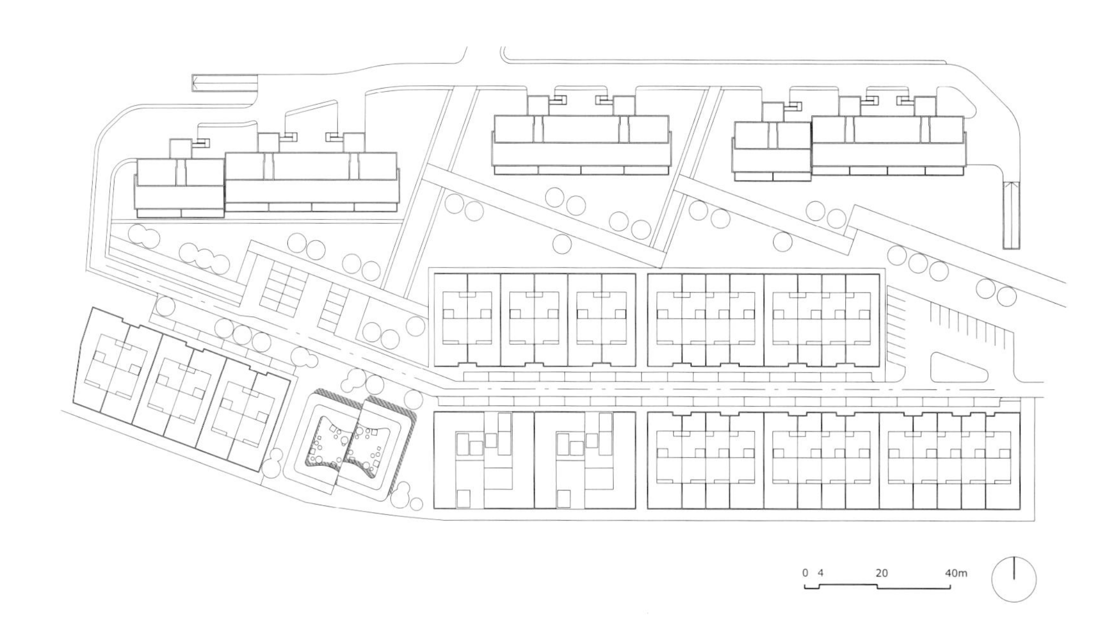

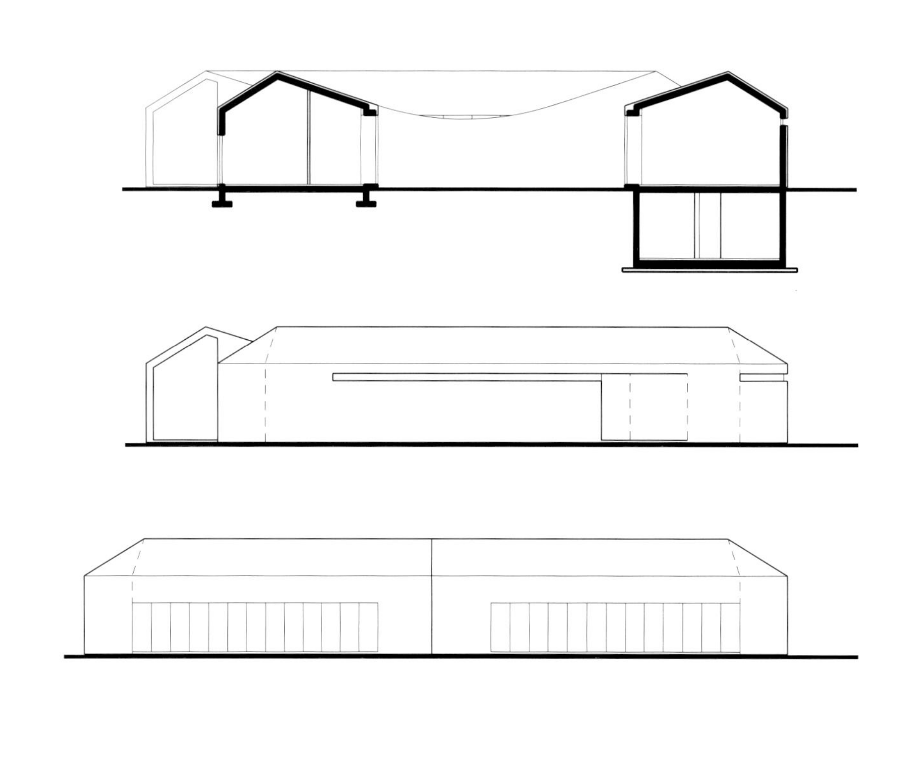

地下一层平面 / The underground floor plan
首层平面 / The 1st floor plan
外景 / Exterior view

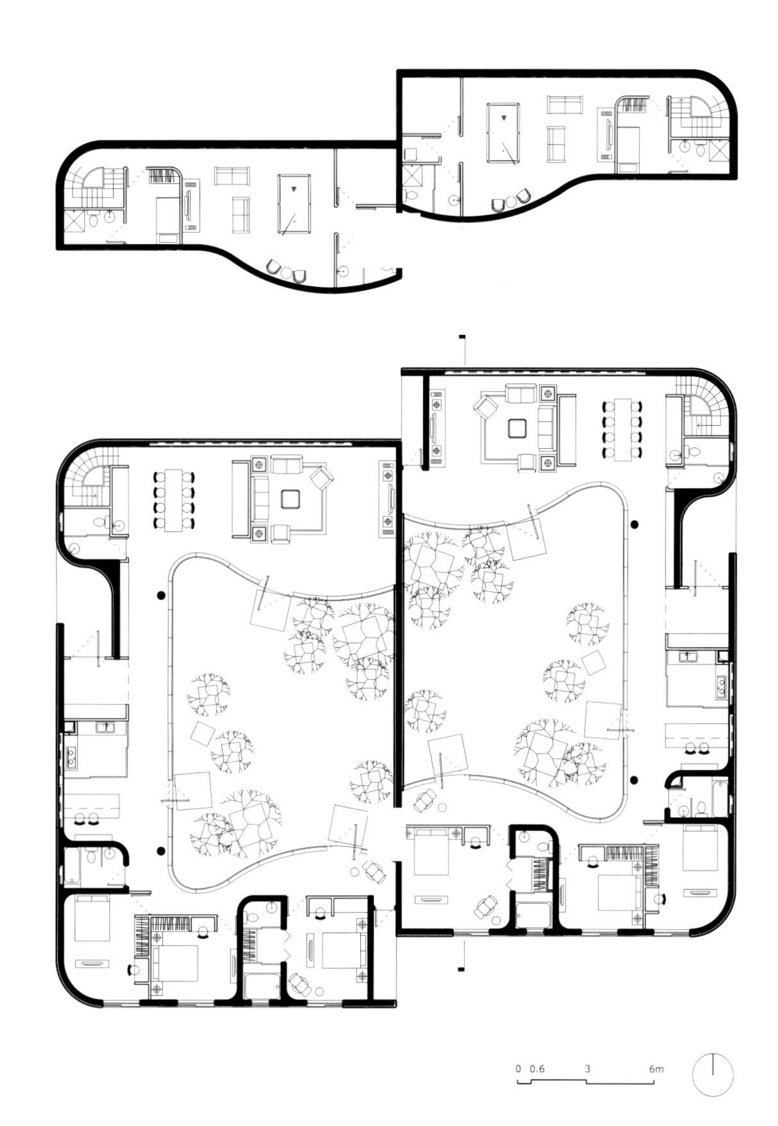

内景 / Interior view

胡越访谈

Interview

采访人／黄元炤
北京　2011.10.13

黄：1995年，您接到一个设计任务，即北京国际金融大厦，是个金融机构的办公建筑，且为您获得业界的声誉。就我的观察，这是个尊重当地城市既有街廓与肌理的设计，由于长安街南北两侧已被分割成大小相近的长方形地块，这些地块皆建有大型的公共建筑，所以，这个项目也恰如其分地以规矩方整的体块出现，而体块与体块之间也形成虚实相互扣合的关系。这部分您有何看法？

胡：以现在的眼光来看，当时我还是有点糊里糊涂的感觉。虽然不是特别年轻，但是也还没有什么太多经验，因为在北京国际金融大厦之前，没有特别多的项目可以做。实际上这个项目业主当时找了一些境外单位做过方案，但一直不太满意。最后，委托到了我们院里来做。当时找的是何总，因为何总原来在二所工作，就找了二所几个年轻的建筑师来做方案，没想到我做的方案被甲方看中了。

黄：北京国际金融大厦，我比较关注的是，在立面上，您利用现代的建筑材料铝合金窗式玻璃幕墙，去表现出传统民族图案的组合。在满足功能的考量上，允许了局部的传统装饰性，即传统在现代的基础上寄居，这其实也回应了后现代主义在中国的其中一条路线，以民族形式的构件与装饰手法去重现民族特色与精神。所以，您从早期参与设计的奥体中心体育馆与英东游泳馆的具象却又抽象的符号、隐喻的象征，到北京国际金融大厦的民族形式的构件与装饰手法，从关注大的形式转到关注小的局部，但仍然在功能考量的基础上，这样的转变，您有何看法？

胡：北京国际金融大厦主要受到两方面的影响。第一方面，因为做英东游泳馆和北京国际金融大厦的时候，当时北京的市委书记是陈希同，他有一个口号叫"夺回古都风貌"，因古都风貌已经没有了，所以，所有北京新建的重点项目，必须得戴上一个琉璃瓦的帽子。其实，当时所有的职业建筑师都对此特别反感，觉得四不像。我在刚开始做金融大厦的时候，绝对不想搞这种戴帽子的把戏，但当时也估计到了，在长安街上设计房子肯定要经过市里的高层领导审批，怕太现代了审不过去，所以当时想到，是不是在细节上做一做这个东西。另一方面，就是努维尔做的巴黎阿拉伯中心，这个建筑在玻璃幕墙上的做法给了我很多启示。

黄：所以，一方面是受到国内政治方面的高层领导审批的影响，一方面可能是受到国外建筑师的设计概念与处理手法的影响。

胡：对。以我个人的考虑，其实我不太赞成那种带有民族符号的煽情的地域主义。

黄：好。就我的了解，1996年的春节，您一个人在设计院里面度过，画图赶工，因为甲方要求需赶紧交图，之后您为了这个项目经常跑工地，一周跑三次。能谈谈当时从设计到图面绘制，再到跑工地，给了您一个什么样的工作经验？这样的过程是否对您以后的工作价值观起到了影响？

胡：北京国际金融大厦是20世纪90年代后期的作品，已经快要接近21世纪。做这个项目当中，不光是职业技巧方面的转折点，也是我思想上一个重要的转折点。在这中间我出过一次国，去了美国，这是我整个人生当中一个重大的转折点。虽然用现在的眼光看北京国际金融大厦，有很多不足的地方，觉得它有好多很幼稚的地方。但是，当时那个时代，从整个专业历程来说，北京国际金融大厦是一个特别重要的节点，对我是全面影响的，不光是专业上的。

黄：1998年建成的北京国际金融大厦设计中，您使用到了独特的窗式幕墙系统，将幕墙分解成4米x3.6米的单元，采用传统方法将其固定在钢框架中，在中心大厅也采用点式连接钻石型玻璃锥顶技术。然后在 2004 年建成的望京科技园二期，您根据体形的变化与功能的要求，产生了 4 种玻璃幕墙，且幕墙的玻璃也分别选用了 Low-e 中空透明玻璃和 Low-e 中空印刷玻璃两种。所以，从幕墙形式的组合到印刷玻璃的运用，您能谈谈对于幕墙的使用吗？当时是否就兴起对幕墙的研究？

胡：我是这样想的，其实在做北京国际金融大厦过程中，出国看了很多建筑，做了好多思考，当时觉得国外有一些建筑师，他的建筑创作是以材料为出发点的。比如安藤忠雄，他做清水混凝土。我就觉得中国建筑师在这方面有个缺失，对材料和构造不了解，完全是被动地使用材料，没有把材料和构造当成是一个创作的动力。我想我是不是能够在这方面做个研究，当时对玻璃比较感兴趣，所以就从这儿开始有意识地去做研究，对那些构造做了一些了解。应该说，对玻璃材料和构造在建筑师这个层面上我是比较了解的。因为专门做了这方面的功课，也订了很多这方面的杂志。

黄：您提到对材料和构造感兴趣，并特别对玻璃做了研究与了解，那我站在一个材料的视点来问。就我的观察，在操作项目时，建筑师有些会从单纯试验材料的设计观点出发，以材料思考带出空间与建筑的走向，材料的选择是影响其作品最终的表现形式。拿日本当代建筑界来说，就有一批建筑师是这样表述着。如安藤忠雄以清水混凝土营造出自然冥想的禅意巨大空间；藤森照信，他独钟情于运用一些自然材料进行创作；坂茂，他让纸成为建筑的一部分，创造出新的建筑构造体；远藤秀平的波形钢板去探索廉价平民材料的可塑性。而我观察到您似乎也是属于此类走向的建筑师。在望京科技园二期，您采用了密肋玻璃与全透明幕墙及低辐射印刷玻璃，使几何量体之间的关系更清晰及对位。在上海青浦体育馆、训练馆改造中，您用聚碳酸酯板材，以纵向与横向的编织手法围绕在建筑物的外层。在上海世博会 UBPA 办公楼，各立面均采用均质开窗的方式，外墙涂料颜色变化、拼接，且由于功能需求形成不同材质（织物膜、铝合金）的体量穿插。从这几个项目观察，您似乎已将视点关注到材料的使用与再研究、再开发上。您自己怎么看待您是属于此类走向的建筑师？

胡：刚才，我也跟你说了，我觉得中国建筑师有点缺失，所以我做了一些知识积累，不能说研究，在实践当中也做了探索。用玻璃我是挺失败的，坦白地说，我在望京科技园二期还有五棵松体育馆都大量地使用了玻璃，积累了一些经验后，现在我不太用玻璃了，因为，我觉得它不太好把握，特别像中国这种施工状态与工期以及和甲方之间的关系，你做不了特别好的实验。你想做实验的时候，加工订货都还没有定呢，所以做不了，等加工订货后，实验也没法做了。这个就不像国外那样可以反复做实验，我觉得玻璃必须到现场反复做实验，但这点很难实现，所以我现在做玻璃，特别慎重，不像原来那么鲁莽了。

另外玻璃的节能环保也有点问题，所以后来我做玻璃是做得比较慎重的。我最近做清水混凝土与木头的多了点，所以我觉得可能向这个方面偏了一些。我觉得建筑设计的构想必须有一个着力点，它包括了很多内容，从社会方面的到专业方面的。材料只是我整个一个大的想法当中的一个部分，可能还有更大的范围需要去考虑，现在我正在做这个工作。

黄：就我的观察，在上海青浦体育馆、训练馆改造中，您在既有的结构体外，新增一层外皮，同时以三种材料相互展现，上层是纵向与横向的编织手法围绕在建筑物的外层，下层是铝合金的穿孔板与铝合金方管的组配。所以，您在对材料的忠实表述下，企图创造出了一条偏向于表皮表象的设计路线，而您在表象这条线上也一直在转变，是一个从硬到软的过程，从玻璃的硬到聚碳酸酯板材的软，也因玻璃是一个整体性的材料，无法编织，只有物件薄且接近于软性的材料如聚碳酸酯板材，才能编织。这部分您自己如何看待？

胡：我还不太同意你说的所谓表皮表象的设计路线。首先，我所处的设计环境并不支持我在一段时间内沿着某一种路线设计，也就是说社会环境不支持我形成某种风格，特别是在国有大院中。因为我经常会接触一些比较大的项目，而大的项目往往建筑师的话语权会很小，因此很难在不同的项目中采用相似的手法。可能有些人不同意我的观点，但我认为形成个人风格除了建筑师个人的因素外，更重要的是社会环境，我觉得我们有时候更像是全科大夫，而不是专科医生。当然我的另一个想法是每一个设计路线都是根据每一个项目的具体情况提出的。

比如说，我最典型的表皮设计就是上海青浦体育馆、训练馆。鄂尔多斯项目，还有上海世博会 UBPA 办公楼，包括我现在刚做好的北京建筑工程学院的房子就不是这样了。比如说，上海青浦体育馆、训练馆，它就需要外面罩一个罩子，我就给他罩一个罩子。上海世博会 UBPA 办公楼，它不需要罩这个罩子，我就不给他罩。我就是这么一个简单的想法，我并不想在所有房子上都做同样一件事，同样一个表皮的设计。

黄：您说的根据每一个项目采取不同的措施，就是建筑师个人对于设计都有不同的一个参照面。

胡：我是根据具体的状态去想这个设计的。

黄：那我们回归到建筑史的视点，通常读建筑史会认识到许多流派或流向，而观察您的作品，若客观定位的话，您近期是沿着一条表象表皮的设计路线在操作，而目前中国当代建筑界确实有部分建筑师关注到表象表皮这一元素，且路线明显。您怎么看您与这些建筑师在表皮运用上的差异性？

胡：也许我的话得罪人。我是这样想的，我还是原来的观点，中国建筑的整体水平，还是低层次的，我们总是要往上看，所以我觉得不管是哪种东西，总是跟现代发达的建筑文化里的某些东西很像，这个是我的观点。在我的整个成长历程中也能反应出这样状态，从原来的 KPF 风格的金融大厦，然后到科技园的风格，其实都跟当时的发达的建筑文化联系着，其实这也是我比较痛苦的一个方面，不光是我的问题，有些人也有同感，但有些人找到了自己的东西。我感觉特别是在建筑的形态上，不管你把它说得多么深奥，最后反应还是要落到建筑的形态上。人家看你，评论你，给你归类也是通过这个形态来归类的，你不愿意这样被归类，人家也给你归到那儿，古今中外都是这样的。

总之，我是觉得我一直是一个比较时尚，且跟风跟得比较紧的人。在设计望京科技园的时候，是 1998 年，当时玩大悬挑、大体块的中国建筑还很少，我有一种感觉这个大悬挑设计挺有意思，所以就去做这件事情了。从我本意上来说，慢慢地我想逃脱这种感觉，但觉得挺难的，因为这就是形象思维的问题。望京科技园有它的设计特点，但它会受到别人形象的严重干扰。所以我是觉得，作为一个中国的建筑师你要向前开往上看的话，是很难的一件事情，总是会不可避免地看到走在前面的人的背影，所以我一直想寻找一条路，能够逃脱这种束缚。当然有人说，这种汇入文化的大潮没有什么不好，从另外一方面来想，这也是可以的。但我是觉得是不是能够逃脱后，形成一个更独特的东西，更有内涵的东西。可是我并不想从地方主义，或是纯情感宣泄的渠道去找，所以，现在我特别关注设计方法，也就是这个原因。

我还是比较关注时尚的，这跟未来有关，因为我觉得不掌握未来的话，我现在没法做，因为建筑是设计，也就是在设计未来，你盖好建筑的时候，肯定是将来的东西，所以，如何在当下想到将来，能让将来站得住脚，这是设计中一个很大的难题。

黄：您愿意跟上潮流，关注时尚。其实，您还是关注到材料方面，一种用材料去体现现代时尚的感觉。

胡：材料，只是我的一部分，我只能承认我原来是比较关注到材料的，我近一段时间，就比较关注到方法了。你可能不知道，我读了个博士，我的论文就是方法论，这是我主攻的研究领域，仔细琢磨了五年，我在这方面有一些想法。

我现在想的方法论，已经到了一个转折点了，我想把方法论变得更深入一点，向未来的发展趋势与社会的需求的方向去转化，因为方法还是建立在最开始的需求上，为了解决问题，才有了方法，所以我现在想找问题，把问题找出来，这是我现在的主要的想法。

黄：您想找问题，似乎回归到一个关注到现实社会层面的思考。

胡：我现在想找问题，然后是材料，最后是专业技巧。我觉得专业技巧肯定需要提高，但是，最基本的东西就是如何采取独特的视角，发现问题，这是我想解决的问题。

黄：其实我有种感觉，您一直不想把自己定性。

胡：对，我觉得应该与时俱进。

黄：您一个项目建成了，您就想要赶紧逃脱开来。

胡：我虽然在大院里，其实我的作品特别少。这是跟我的工作方式和机缘有关系。比如说我做了很多特别大的项目，我难于启齿告诉别人，五棵松体育馆被别人破坏，我也始料未及，没想到会有这么大的干扰，在关键的时候，把形式都给变掉。

我觉得可能每一个大的项目都需要孕育很长时间，这跟大设计院有关系，有特别大的项目一干就是好几年，然后，盖不起来，时间拖得特别长，有的时候，干小项目，四五年就已经干了七八个了，但是，大项目可能一个都没干完呢。

因此我在项目上的变化，可能看着像是每一个项目都有一些变化，其实这跟做大项目有关。对于某些人他做小项目的时候，可能已经有好几个项目了，但是我做大项目一干就是好几年，思想已经有些变了。

黄：您自己有没有个人的设计中心思想或信仰？是不是有在塑造自己的设计语言，还是一些表征？这个问的比较纯粹一点，比较形而上的一个话题。

胡：我估计你可能搜过我原来的论文和发表的文章。有时我都觉得我有了方向，后来发现又没有方向了。

黄：对，跳得非常快。

胡：你现在问我自己的设计思想，我说不出来。

黄：还在摸索或困惑中？

胡：我感觉建筑师是挺自私的人群。搞艺术创作的人都有点这样。他经常想把自己的欲望强加给别人。但是，建筑又不是纯私人表达的载体，所以我觉得一方面你不应该一味地去迎合一些人的想法，因为你是一个专业人士，你要引领别人，另外你也不能只注重个人思想的表达。现在最大的问题不是什么很高深的想法，而是设计的建筑要给人感觉舒服，这是一个最终的理想。就是人在那儿待着的时候，是很舒服的感觉，不管是室内，还是建筑和周边的建筑所形成的室外空间，要给人感觉很舒服、很轻松，我觉得这是一个终极的目标。而不是要一个吓人一跳，特别怪异，或是出了一个什么招数。

我为什么有这么一个想法，在年轻的时候大家都喜欢跟风潮的，特别崇拜一些建筑师，可是后来我发现到现场一看，那些房子挺恶心人的，跟由媒体所传达的信息完全是两码事，尤其是现在很时尚的那些建筑师，我到现场去看他们的建筑，都给我特别大的失落感，怎么做成这样，看着真难受，跟古典时期留下的普通房子比起来，有特别大的反差。如果建筑，它有社会的使命与历史的使命的话，我感觉可能更应该是那个给人感觉很舒服的建筑，而不应该仅仅是一种时尚。

当然，建筑应该反映时代的面貌，我感觉先进的东西肯定在时尚里头的。但这些时尚，可能最后大部分都泥沙俱下，大浪淘沙，都成为垃圾，被历史证明是垃圾。但是肯定有一两个是外在历史前端的时尚的东西，又能给当下的人和后代的人以舒服的感觉，这当然是最高境界。虽然我不一定能达到这个境界，但至少这是一个愿景。

我觉得风格，都是一个次要的东西。你所设计的房子，要让人多年以后去看了，感觉仍挺舒服，那才是最主要的。我认为室外给人舒服比室内更重要，因为我觉得室外是为大多数人服务的。一个房子，你进去的机会很少，只能是那房子的主人，或者在里头居住或工作的人，但是，大部分人都是在它所形成的城市公共空间里穿行，如果这个城市公共空间给人感觉很难受的话，那你就是一个罪人。

建筑是一个很重要的因素，因为它是城市公共空间的背景。建筑师设计了这个背景，所以我觉得这是建筑师最大的使命，应该从这个角度看这件事情，剩下的那些，比如如何去判断未来的发展，如何引领时尚，这是第二个层面，我也想追求这些东西，但是，我觉得这些不是最高层面的追求。

黄：您说的室外比室内更重要，关注到的就是城市公共空间的营造与社会责任。谈到责任，我想到一点，就我个人对于中国建筑发展的研究，您们这一代的建筑师，由于讯息的缺乏，所以很难全面地了解到以前所发生的建筑史实，或者只知道片断与零碎的记录。而您所处的 20 世纪 80 年代、90 年代中，传统与现代一直是被争论的话题。但是传统与现代的争论早从 20 世纪 20 年代、30 年代就开始了，所以，您怎么看待现今的中国现代建筑？那传统建筑又该如何自处于现今的社会上？

胡：我考虑过这个问题。首先，先抛开这个题目，从文化、传统、地域性来考虑这个问题。我对地域性做过一点思考，我采用唯物主义的观点，可以认为先由物质来决定精神层面的东西，为什么在中国形成这样的房子，在希腊形成那样的房子，是有机缘巧合的，但也有很重要的因素，就是当地的气候、材料和文化的传统而形成这些房子。那为什么在古代形成了这么多色彩斑斓、变化万千的文化？我觉得是由于当时的交通，联系不发达而导致的隔阂才出现这样的情况。动物的进化也是这样，澳大利亚的动物跟欧亚大陆的动物完全不一样，就因为彼此间不联系，所以就形成了很独特的东西。

那么现在的世界不存在这种隔阂了，还有差异存在主要来自古老文化的惯性，那么在目前的情况下，地域的差异性很难再产生。我举个例子，比如过去北欧的房子，坡顶都特别大，是因为防雪，现在的技术一个平顶全可以解决，所以，我是觉得从物质基础的层面上看，形成一个地方文化

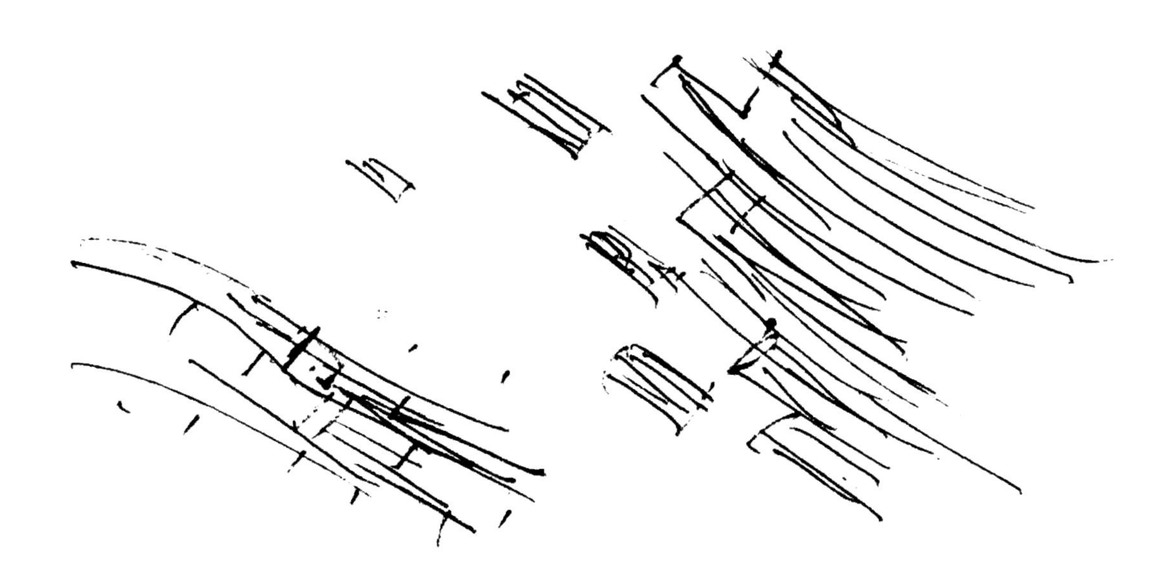

的这种建筑不太可能以单纯的形式来表现，可能更多的表现在社会需求上。

我碰见过一个中国联通的官员，他说中国有很多互联网的技术，已经在世界上处于顶端了，为什么，就是服务的人群太大了。在国外，可能几百万就是很了不起的数字了，咱们这儿几千万，都不算一个大数字。要去解决这么大数量的问题的时候，就会出现一个新的技术。所以，我觉得从中国来说，有中国特点的东西不是对过去形式的重现，因为过去那个已失去存在的基础，物质基础没有了，因为一个技术全能完全解决问题。因此解决中国自己的问题变得非常重要，我是不是能够看到中国自己的问题，我觉得这是关键问题，所以依我看，方案目标是找一个新的视点去看中国特有的问题。当然现在中国的情况，我觉得文化的惯性还很大，传统的文化起着很大的作用。但是，建筑师不应该刻意地再回到过去，过去除了在感情上的一点感觉之外，除了煽情之外，其他都没有任何意义了。在这种情况下，中国对待自己文化的态度，应该是把中国古代的东西好好保护好，现在保护得太不好了，然后，就好好做个新的房子。你看看建筑文化发达的国家，也有很悠久的历史，比如法国、英国，但他们的建筑师从来不往后看。

我觉得中国现在的状态，一个就是要好好地保护古代传统的东西，认真地去做新的建筑，另外一个，在文化发展当中应该努力地缩短跟先进文化之间的距离，我觉得只有超越这个东西，达到了先进文化层面，你才会发挥出真正自己独特的东西。因为到那个时候，你往上看的时候，只有天，没有别人了，才能够发展成自己的东西。所以，现在急于发展自己的地域性，我觉得没有意义，我的观点就是这样，就是缩小跟先进文化之间的距离，这是咱们这一代人的努力方向，而不是说从古代里弄点花样，我觉得那个也可以满足有些人的心理需求，但是我觉得那不是大方向，那是主菜里添的那点佐料，可以那么做，我不反对，但是我觉得大方向不是那个路子。

黄：您觉得中国现代建筑应该要缩短跟最先进文化之间的差距，这个缩短差距，就像您说的您要跟上潮流，与时俱进，所以，您想让您的建筑处于一个进行时、未来时，而不是过去时，但对于传统，所谓的过去式，就要好好地保存它。

胡：对，没错，传统要好好保存，好好学习，从传统中汲取养分，但是提取的养料不是为了宣传用的。

黄：好，设计过程中，您是如何产生的想法与构想的？

胡：我比较关心的项目，我都自己再写一个任务书，建筑师任务书，我觉得这个也是这几年比较关注的内容，就是我要从中发现独特的问题，因为我觉得建筑的问题，可能有无数个问题都摆在那儿了，你可以去发现它，但是发现哪个问题，哪个问题作为创作的重点，这是个人的价值判断，我是想努力地发现一些独特的问题，提出独特的解决方法，然后通过这些方法来表达建筑并形成一个独特的建筑。我是这么想的。

作品年表

Chronology of Works

★——收录作品 ●——实现作品 ○——未实现作品

建筑名称 北京国际金融大厦 ★●
设计团队 胡越、汪安华、苑泉、齐五辉、于永明、赵福田、胡又新、甘虹
摄影 付兴
建设地点 北京市复兴门内大街闹市口
设计时间 1996 年
竣工时间 1998 年 5 月
用地面积 1.8 公顷
建筑面积 103313 平方米
获奖情况 第二届詹天佑土木工程大奖
全国第九届优秀工程设计金奖

建筑名称 望京科技园二期 ★●
设计团队 胡越、郃方晴、王婷、冯颖玫、王皖兵、于永明、胡又新、白冬、孟秀芬、甘虹
摄影 陈溯、杨超英
建设地点 北京市朝阳区望京新兴产业区北部
设计时间 2000 年
竣工时间 2004 年 4 月
用地面积 25916 平方米
建筑面积 46297 平方米
获奖情况 第三届中国建筑学会建筑创作佳作奖
WA 中国建筑奖优秀奖
2006 年度全国优秀工程设计银奖

建筑名称　五棵松棒球场 ★●
设计团队　胡越、顾永辉、孟岭、马学聪、齐五辉、韩永康、王雪生、范珑、张野、申伟、胡又新
摄影　杨超英
建设地点　北京市海淀区五棵松
设计时间　2005 年
竣工时间　2007 年 7 月
用地面积　12 公顷
建筑面积　14360 平方米

建筑名称　上海青浦体育馆．训练馆改造 ★●
设计团队　胡越、邰方晴、缪波、牛敏婕、张燕平、薛沙舟、申伟、李辉
摄影　付兴
建设地点　上海市青浦区体育场路
设计时间　2007 年
竣工时间　2008 年 2 月
用地面积　5536 平方米
建筑面积　8100 平方米
获奖情况　北京市第十四届优秀工程设计三等奖

建筑名称　上海 2010 世博会 UBPA 办公楼改造 ★●
设计团队　胡越、邰方晴、孟岭、冯婧萱、范波、魏宇、陈蕾、任重、张谦、才喆、杜鹏
摄影　邵峰、付兴
建设地点　黄埔区南车站路
设计时间　2008 ~ 2009 年
竣工时间　2010 年 4 月
用地面积　1562 平方米
建筑面积　6052 平方米
获奖情况　2011 年度全国勘察设计行业优秀工程（世博工程）三等奖

建筑名称　北京建筑工程学院新校区学生综合服务楼 ★●
设计团队　胡越、邰方晴、张晓茜、张俏、唐强、田新潮、程春辉、王旭
摄影　陈溯
建设地点　北京市大兴区
设计时间　2010 年
竣工时间　2011 年 12 月
用地面积　5567.31 平方米
建筑面积　4443.19 平方米

建筑名称　鄂尔多斯 20+10 项目 D2 地块 ★●
设计团队　胡越、郜方晴、曹阳、吕超、刘全、赵默超、项曦、陈彬磊、李婷、李志武、陈栋、薛沙舟、李丹、张永利
建设地点　鄂尔多斯东胜区
设计时间　2011 年
竣工时间　在建
用地面积　8412.5 平方米
建筑面积　24197.58 平方米

建筑名称　杭州奥体中心体育游泳馆 ★●
设计团队　胡越、顾永辉、游亚鹏、曹阳、于春辉、孟峙、缪波、冯婧萱、沈莉、张燕平、李国强、马洪步、徐宏庆、郑克白、祁峰、张成、刘晓茹、胡又新、张永利、吴威、景蜀北
建设地点　杭州萧山区
设计时间　2009 ~ 2011 年
竣工时间　在建
用地面积　227900 平方米
建筑面积　396950 平方米

建筑名称　延庆小别墅 ★○
设计团队　胡越、刘亚东
建设地点　北京市延庆县妫河创意园区
设计时间　2010 年
竣工时间　未建
用地面积　723 平方米
建筑面积　416.5 平方米

建筑名称　平谷 3 号住宅 ★●
设计团队　胡越、刘亚东
建设地点　北京市平谷区马坊镇
设计时间　2011 年
竣工时间　在建
用地面积　702 平方米
建筑面积　476 平方米

建筑名称	五棵松体育馆 ●
设计团队	胡越、顾永辉、邰方晴、齐五辉、范珑、胡又新、沈莉、薛沙舟、甘虹、罗靖、陈莉、申伟、高峰、闫峰、张燕平、游亚鹏、孟峙、柳颖秋、陈盛、张永利
摄影	付兴、杨超英
建设地点	北京市海淀区五棵松
设计时间	2005 年
竣工时间	2008 年 1 月
用地面积	52 公顷
建筑面积	63000 平方米
获奖情况	第五届中国建筑学会建筑创作优秀奖 第八届中国土木工程詹天佑奖"建筑工程奖" 2008 年度全国优秀工程勘察设计金奖

建筑名称	五棵松文化体育中心文化体育及公共服务设施 ●
设计团队	胡越、罗靖、邰方晴、齐五辉、沈莉、范珑、薛沙舟、陈莉、吕静、甘虹、胡又新、申伟
建设地点	北京市海淀区五棵松
设计时间	2007 年
竣工时间	2008 年 12 月
用地面积	218969.79 平方米
建筑面积	334400 平方米

建筑名称	上海世博会亚洲六国馆 ●
设计团队	胡越、顾永辉、缪波、范波、张胜、王新、周有娣
摄影	游亚鹏
建设地点	上海，浦东野路，世博园内
设计时间	2008 年
竣工时间	2010 年 3 月
用地面积	6000 平方米
建筑面积	4000 平方米

胡越简介

1986 年毕业于北京建筑工程学院建筑系
2011 年获得清华大学工学博士学位
现为全国勘察设计大师
北京市建筑设计研究院总建筑师
胡越工作室主持建筑师
中国建筑学会理事

其作品获得：
20 世纪 90 年代北京十大建筑
第二届、第八届詹天佑土木工程奖
全国第九届优秀工程设计金奖
2004 年度全国优秀工程设计银奖
2005 年度部级优秀勘察设计一等奖
2006 年度全国优秀工程设计银奖
2008 年度全国优秀工程勘察设计金奖

代表作品有：
北京国际金融大厦
望京科技园二期
五棵松文化体育中心
上海青浦体育馆、训练馆改造
上海 2010 世博会 UBPA 办公楼改造
北京建筑工程学院新校区学生综合服务楼

Profile

Graduated from Department of Architecture, Beijing Institute of Civil Engineering and Architecture in 1986.
Tsinghua University in 2011 with his Ph.D. degree.
Master of National Survey and Design.
Chief Architect of Beijing Institute of Architectural Design.
Principal of Huyue Studio.
Director of Architectural Society of China.

Awards:
The 1990's Beijing Top Ten Construction
Zhan Tianyou Civil Engineering Award in 2002 and 2008
The 9th Gold Award of National Excellent Architecture Design
Silver Award of National Excellent Architecture Design in 2004
Silver Award of National Excellent Architecture Design in 2006
First Prize of Outstanding Survey Design in National Ministry,2005
Golden Award of National Outstanding Engineering Survey Design,2008

Representative Works:
Beijing International Finance Mansion
Wangjing Science & Technology Park Phase II
Wukesong Baseball Field
Shanghai Qingpu Gymnasium-Training Center Rehabilitation
Shanghai Expo 2010 UBPA Office Building Rehabilitation
BUCEA New Campus Student Comprehensive Service Building

致谢

特别感谢建筑师王昀为此系列专辑出版所作出的贡献。感谢卢超同学为本书编排所付出的努力。感谢胡越工作室各位同仁为该书中的各项目顺利完成所付出的努力。感谢我的家人为我所做的一切！

胡 越
2013.1.16

Acknowledgement

My deepest gratitude goes to the architect, Wang Yun, for his contribution to the publish of this series. Special thanks should go to Lu Chao who has put considerable time and effort into the layout design. I also would like to express my sincere gratitude to my colleagues of Huyue Studio for their support on the completion of this book. Finally, I'm indebted to my beloved family for everything they done.

Hu Yue